LIMITLESS MINDS INTERVIEWS WITH MATHEMATICIANS

LIMITLESS MINDS

INTERVIEWS WITH MATHEMATICIANS

ANTHONY BONATO

AMERICAN MATHEMATICAL SOCIETY

Providence, Rhode Island

2010 *Mathematics Subject Classification*. Primary 01A70, 01A65.

The cover photo of Maria Chudnovsky (top row, right) is courtesy of Maria Chudnovsky.

For additional information and updates on this book, visit
www.ams.org/bookpages/mbk-118

Library of Congress Cataloging-in-Publication Data

Names: Bonato, Anthony, 1971- author.
Title: Limitless minds : interviews with mathematicians / Anthony Bonato.
Description: Providence, Rhode Island : American Mathematical Society, [2018]
   | Includes index.
Identifiers: LCCN 2018034281 | ISBN 9781470447915 (alk. paper)
Subjects: LCSH: Mathematicians–Interviews. | Mathematicians–Intellectual life.
Classification: LCC QA28 .B66 2018 | DDC 510.92/2–dc23
LC record available at https://lccn.loc.gov/2018034281

Contents

Preface

I think back to the formative moments in my youth when I became excited by mathematics. As a young teen in my 10th grade mathematics class, I recall our teacher assigning proofs of trigonometric identities as homework. I was bored with the assigned questions, so I tackled the toughest one in the text, which had a star next to it (which meant it was extra tough). The next day, the students were asked to present our homework on the board voluntarily, and I presented the solution to the starred exercise. While I wrote on the board, our math teacher said: "Oh, that one."

Every mathematician is a person with a story. Each of them has memories from childhood or their teen years when things clicked for them mathematically. Many of them had teachers or family members who helped them progress in their academic career; some, however, had people discourage them from studying mathematics and persevered.

My goal when writing this book was to give a snapshot of exceptional mathematicians and to tell their stories through interviews. I set out to answer some key questions: Who and what were the influences that pointed them towards mathematics? Why do mathematicians devote their lives to discovering new mathematics? How do they see mathematics evolving in the future?

These are some of the questions I address with interviews from twelve of the world's leading mathematicians. These interviews were originally featured on my blog *The Intrepid Mathematician*. The interviews are popular features on my blog, which has received a total of 120,000 views from 75,000 unique visitors. The blog has been a labor of love. As evidence of its impact, in 2017 the blog was nominated for Science Borealis 2017 People's Choice Awards: Canada's Favorite Science Online.

I hope that the book will enlighten and inspire readers about the lives, passions, and discoveries of mathematicians. Maybe you are a mathematician yourself, or a student of mathematics. Or maybe you are a scientist or working in STEM, or simply a person curious about what makes mathematicians tick. This then is the book for you. The book is about mathematicians and mathematics, but it is written in accessible language, with only a limited amount of technical terms or formulas.

Experience tells me that readers would prefer to buy a physical book rather than click through a website. There is something special, talismanic even, about holding a book in your hands as you are now.

Why choose mathematicians as subjects? What about interviewing physicists or medical doctors or engineers? My response is another question: "Why not?" There is, after all, a growing interest in mathematicians in popular culture witnessed, for example, by the success of books and films such as *Hidden Figures*, *The Imitation Game*, and *The Man Who Knew Infinity*.

No other book has the exact premise of this one, although I recommend the books of Albers and Alexanderson, *Fascinating Mathematical People: Interviews and Memoirs* and *Mathematical People: Profiles and Interviews 2nd Edition*, published in 2011 and 2008, respectively. Both books include interviews, biographical text, and memoirs of mathematicians both living and dead. In contrast, this book has recent interviews from a broad swath of the current mathematical community.

Each of the twelve interviewees has their own unique story. I chose subjects with a view towards inclusion, diversity (both gender diversity and other kinds), and topic matter. Think of these dozen subjects as a peek into my catalogue of living mathematical heroines and heroes. Each interview is a chapter, with brief opening text discussing the subject, along with biographical data and personal stories of mine about them. There are photos of the subjects and other images as well.

I do hope you enjoy the interviews. If you come away from reading the book with any insights into the limitless imaginations of mathematicians, then I've done my job well enough.

I would like to thank Ina Mette, Marcia Almeida, and the American Mathematical Society for their enthusiastic support of the project. My first two books were with the AMS, and it is always a pleasure to work with them. I'd like to thank all the dozen interview subjects for sharing their stories. A special thank you to the unwavering support of my spouse Doug, whose love itself is limitless.

I dedicate this book to all of my graduate students, past, present, and future. May the book guide, inspire, and uplift you.

Anthony Bonato

Chapter 1

Interview with Alejandro Adem

Alejandro Adem is the Canada Research Chair in Algebraic Topology at the University of British Columbia (UBC). Alejandro was born in Mexico City and completed his doctorate at Princeton University. Not only is he a stellar mathematician, but he is the CEO and Scientific Director of Mitacs Canada. Mitacs is a not-for-profit organization which began as a federally funded Network of Centres of Excellence, and it is devoted to supporting research and training students and post-docs through partnership grants with industry.

Alejandro is an award-winning mathematician. A few of his honors include the Jeffery-Williams Prize from the Canadian Mathematical Society, an appointment as a Fellow of the American Mathematical Society, and a Vilas Associates Award from the University of Wisconsin-Madison. He is also the managing editor of the highly impactful journal *Transactions of the American Mathematical Society*.

My chat with Alejandro was inspiring, and I found him so open and generous with his answers. The breadth of his experience and knowledge was impressive. He talks about the K-theory of orbifold cohomology in one part of the interview and then Canada's innovation ecosystem in the next.

This interview was conducted in February 2017.

Figure 1. Alejandro Adem. (Photo licensed under Creative Commons Attribution-ShareAlike 4.0 International (CC BY-SA 4.0).)

∞

AB: What were your early mathematical influences? Did anyone, before your studies at university, inspire you?

AA: I'm in a special situation because both my father and my uncle were mathematical scientists. My father had a Ph.D. in Applied Mathematics from Brown University, and my uncle had a Ph.D. in Mathematics from Princeton University. They were both very distinguished scientists in Mexico, where I was born. They were the first generation on that side of the family to go to college. They were always inspiring examples to me.

On the other hand, teachers inspire much of what we do in mathematics. I did have some inspirational teachers, especially one in middle school who used the "new math" which I found fascinating. He taught me a love for the conceptual aspects of mathematics that carried me through for several years. When I was choosing career options, mathematics was on the table because of my family background, and my teachers motivated me. So, going into the field was a natural decision for me.

When I was in school, I realized that if you understood the underlying method in mathematics, then you didn't have to memorize facts like in other subjects such as history or biology. In those subjects, at least at that time, there was a huge emphasis on knowledge acquisition and memorization. While in mathematics, if you understood something, you could remember it. For me, it is the same in life: if you understand the process, you can remember it. For me, it was powerful. It unlocked a part of my brain, one that was not necessarily conscious.

AB: How did you come to work on your doctorate at Princeton University and what was your experience like there?

AA: I did my undergraduate at the National University of Mexico, and when I got to Princeton it was rather intimidating. My classmates included top-ranked students from a top place in the US. My main objective was to listen and learn. At Princeton, you learn as much from your classmates as you do from your professors. There are so many brilliant people there.

However, you can also understand what it means to have mathematical insight, which you have to separate from sheer intellectual power. It is great if you have both. There were individuals there who had great intellectual power but little insight. They weren't natural mathematicians. It was a revelation for me when I held my own and did well, based on mathematical intuition and hard work.

Figure 2. Princeton University. (Photo from Shutterstock.com.)

We had very inspiring teachers there, including my advisor Bill Browder. The professors were gentle, and they unlocked the mysteries of higher mathematics for us in a nice way. There were no grades in the courses. It was all about learning and doing your research and supporting you through that process.

AB: How did you get to work with Browder?

AA: They have a tradition there of tea time every day around 3:30 pm. The professors would come down to the common room, and the protocol back then was that you would wait for them to come to tea and then approach them. I remember meeting him there around the time of my qualifying exams, and he was very friendly. He remembered my application to the program and invited me to his office to talk mathematics. We discussed things I could work on, and I went on my own to study the topics he gave me.

Figure 3. William Browder. (Photo author: Konrad Jacobs. Photo source: Archives of the Mathematisches Forschungsinstitut Oberwolfach.)

AB: You are an expert in algebraic topology and the cohomology of groups. Would you explain what these areas are in simple language?

AA: Algebraic topology is a part of mathematics that tries to understand intrinsic geometric properties of objects that appear in the mathematical world, for example in physics, and in everyday life. We may try to measure, for example, the number of holes in an object, or identify if one object can be deformed into another. As you know, there is a famous analogy between the donut and the coffee cup.

In areas of physics, we may naturally encounter objects arising from topology and geometry, so topology is a fertile ground for connecting higher mathematics and ideas from the natural sciences. We also study objects such as knots, and it can be seen for example that the knotting of DNA connects topology with biology. There are also interesting applications of topology in economics.

We can separate out algebraic invariants that will determine these fuzzier, more complicated geometric structures. If you can derive a number that tells you that you can deform one object to another, then you are in great shape. The most basic example of that is the Euler characteristic, which many times can characterize a space. For example, for orientable surfaces, that number contains a lot of information. You can derive even finer invariants. These algebraic and topological invariants can, in many cases, help describe the original space and the geometric problems that it encodes.

Cohomology of groups is a bit more technical and harder to explain, but the idea is to move from groups to topological spaces. There are nice correspondences, which assign to a particular group a space with geometry that fleshes out properties of the group. There is a rich direction there called geometric group theory which is quite important now. The spaces built out of group theory have a much greater impact than you would expect. Cohomology of groups computes invariants associated to these spaces, which are quite general. You can go from a group to a space, and then compute algebraic invariants of it (in this case, the cohomology ring). So, it becomes a fine science in its own right, with fantastic contributions from people like Daniel Quillen who laid the foundations of the subject for finite groups. There are many applications of group cohomology which connect areas of algebra and topology in unexpected ways.

AB: What are the research directions you are working on right now?

AA: One of the basic topics I have worked on in my career is topological symmetries or group actions. Given a manifold, can you describe or characterize groups that act on the space in a suitable way, such as free actions (that is, actions without fixed points)? There is a classical problem that topologists solved, which is understanding the finite groups acting freely on a sphere; that is, the so-called Spherical Space Form problem. That leads to many interesting questions surrounding the groups and the algebraic topology involved. Many sophisticated equivariant techniques have been developed to study these questions. With more complex groups and structures, the problems become quite challenging.

One of the most interesting results I proved (joint with Jeff Smith) was characterizing spaces with periodic cohomology: one that repeats itself in sufficiently high degrees. That can be described very effectively using methods from algebraic topology modelled on techniques in group cohomology.

I've also worked on topics arising from physics, and that led to very fruitful collaborations; for example, with my former colleague Yongbin Ruan who works in symplectic geometry and mathematical physics. We wrote a book on orbifolds and developed models for K-theoretic versions of orbifold cohomology. That ties into questions about resolutions occurring in string theory.

Through that, I also became interested in mathematics arising from Lie groups, such as structures associated with their commuting elements. With my collaborators, we developed commutative K-theory, which comes from the commuting elements in the unitary groups. There are also infinite loop spaces involved.

In my work, there is usually a group involved, symmetries, and functors from a group to a space. There are very sophisticated tools available to us in algebraic topology. But to me what is most important is to have an output that can be appreciated by a mathematician at large.

AB: You were at PIMS [the Pacific Institute for the Mathematical Sciences] as Deputy Director from 2005 to 2008 and then as Director from 2008 to 2015. Would you tell us about PIMS and how you came to those positions?

AA: I was a professor at the University of Wisconsin, and I saw an ad for the Deputy Director at PIMS and also one for a Canada Research Chair at UBC. I sent an e-mail inquiring. I always liked Canada and BC, as my father would spend a lot of time here in the summers. It was a dream of mine to live here. They were interested and wanted me to send my CV. It was fascinating to meet with my colleagues here as well as with the then director at PIMS who was Ivar Ekeland (who is a very inspiring figure). What I liked about my interview was that we talked about the Serre spectral sequence, and not about bureaucratic detail. He wanted to know about it! I was impressed by the mathematical level of PIMS. I took both positions. I should mention that I knew a lot about mathematical institutes based on my role at MSRI [Mathematical Sciences Research Institute] in Berkeley as Chair of the Scientific Advisory Committee for four years.

I moved to Vancouver with my family, and I love it here in Canada and BC. Working at PIMS was a special experience as it is a consortium of western universities in Canada and Washington state. PIMS was founded under the inspiration and guidance of people like Nassif Ghoussoub, Ed Perkins, and Peter Borwein who are all important Canadian mathematicians.

Part of my job was to visit institutions in the consortium and help create a critical mass of research strength that would rival concentrations in places like Toronto and Montreal, which have excellent strength in mathematics. PIMS is a grassroots organization, including collaborative research groups that tie together researchers at different universities to work on central problems in their field and also directly impact on the training of personnel. It was a great pleasure to work with my colleagues in western Canada and a big responsibility. There are always challenges with the changing nature of funding mechanisms in Canada.

AB: You're in your second year as CEO and Scientific Director of Mitacs. What are the aims of Mitacs and what is your role there?

AA: I was invited to apply for my present position at Mitacs, which as you may know was established by the three mathematics institutes of Canada: CRM [Centre de Recherches Mathématiques], Fields, and PIMS. It was a product of the imagination and the boldness of the mathematics community and built from the bottom up by my colleague Arvind Gupta.

Mitacs has evolved into an organization that now goes well beyond mathematics. It involves all the disciplines of knowledge. The key thing Mitacs does is connect universities with industry. The biggest program we have is called Accelerate, and it provides internships in companies for Canadian graduate students. Students go to a company for four months and work on a research project with direct economic impact. We have a high-level research committee that adjudicates proposals. The students work on research that will enhance the products and services at the company. Typically, a company has a problem that has to be solved with a view towards commercialization. Industry has to pay fifty percent of the internship. There is a huge buy-in from industry to this program; this past year we have delivered close to 4,000 internship units across the country.

We have people on the ground who make these connections. If you don't have people creating links between professors and industry, then it's almost impossible to do. We are a national organization now, and it has grown quite a bit over the last ten years.

We also offer an international program called Globalink that brings in talented students from a targeted list of countries for research projects in the summers. And we offer scholarships to bring them back as graduate students. The point is to use research in a targeted way to bring the best minds to Canada. Also, we have reciprocal programs to send students abroad—it is important to send Canadian students out as we lag behind other countries such as the USA in this process of global mobility and internationalization.

At Ryerson University, you see all around you that innovation and globalization are coming together. The knowledge-based economy has been going into hyperdrive. We see the Globalink program as a way of connecting the innovation ecosystem in Canada with that in other countries such as China, India, France, Brazil, and Germany, among others. Our programs are very much focused on addressing the challenges faced by this country.

As CEO and Scientific Director, it's been fascinating to learn about a completely different structure than at a university or institute. Mitacs is a not-for-profit company. We have around sixty member universities. We deal with the private sector and governments (both federal and provincial). Our programs are focused on graduate students and provide them with experiential education. My view is that every graduate student should have the opportunity of an

internship; not mandatory or prescriptive, but at least given the opportunity. We have instances of programs that embed our internships
in their curriculum.

Mitacs originally stood for Mathematics of Information Technology and Complex Systems. Now it is just an acronym! I always
remind the staff that the "M" in Mitacs is for mathematics. Mathematics is at the heart of innovation. The more I work in other areas
and talk to companies and government, the more I am convinced that
mathematics is at the center of it all. What we do is of core relevance
to society and everything around us.

Figure 4. Alejandro at Mitacs. (Photo courtesy of Mitacs.)

AB: That is a very powerful message. I speak with many intelli-
gent people who don't understand the central role of mathematics
in science and other disciplines. I asked this next question to Nassif
Ghoussoub when I interviewed him. What do you think is the role of
mathematicians in shaping public policy? For example, on matters
pertaining to public math education and in agencies like Mitacs or
NSERC [Natural Sciences and Engineering Research Council]?

AA: The knowledge that mathematical scientists bring, especially to the handling of data, is highly important. We will naturally be contributing through our work, especially given that we think clearly and logically about data as part of our research. Recently some of the top people in Canada working in artificial intelligence have been recruited to places such as Microsoft and Google, so it's an ongoing challenge for Canada to maintain intellectual leadership. Mathematical scientists form a vital backbone for data science not only through our research in topics such as probability and theory of networks but also through our essential role in the training of university students.

Mathematicians are less involved in the political sphere than they could be. I would urge them to get more involved, like what Nassif, myself, and Arvind Gupta have done. I also understand that mathematics is a quiet affair and we need time to focus on our research. However, I am convinced that we need to be smart about communicating what we do to other scientists and governments, to ensure it receives the attention and support it undoubtedly deserves.

Education is enormously important. Since my kids have been in school, I hear a lot about that. My wife Melania works at UBC doing mathematics outreach. She has run a very successful summer program for indigenous high school students and also works extensively with teachers. Students often fail to learn the mathematical concepts and skills required for success in STEM [science, technology, engineering, and mathematics] disciplines and the mathematics awareness in society is not where we want it to be. It affects the kinds of students we see at university. We talk now about "fake news", but "fake numbers" have been around for much longer. That can lead to poor decision making at all levels.

I would like to see mathematicians involved at the local level, involved with schools. I would like to see our colleagues more involved in the teaching of teachers. Some do it now and do a great job. We many times criticize what is taught in the schools. But have we reached out to schools to provide support or mentorship? For example, UBC Math has a program to send undergraduates to schools and serve as a resource. It's not easy, but it is worthwhile, and we have a big role to play.

AB: You are active on social media on both Facebook and Twitter. What do you think is the importance of social media for mathematicians in the 21st century?

AA: Absolutely it is important. Kids live and die by social media! They are more likely to watch a YouTube video on a topic than pick up a book. Social media is their first point of contact.

Some mathematicians like Ken Ono and Jordan Ellenberg already do that, and you do an excellent job of that. Social media can be more than your personal view or political view. If there is a source connecting people on social media to mathematical ideas, opening up people's minds, then that can be very influential. Things you pick up on Twitter or Facebook can plant a seed in your mind which afterwards can blossom in an unexpected way.

I hope we have an expanding presence on social media with mathematicians. A mathematics and media program or training our colleagues on that would be interesting. At Mitacs, we have recently launched a program called Science Policy Fellows where we seek to embed academics in government departments. That is a way to send academics to influence policy. Mathematicians applying for that could create opportunities and make new connections.

We have to create conditions and programs to empower people to do original research and connect mathematics with society. It's like a mathematics problem: we have to use all tools at our disposal.

AB: What advice would you give to young people thinking about studying mathematics?

AA: I think that when you study mathematics, you need to bite the bullet and take the hard courses. You have to understand what mathematics is about and the incredible knowledge there. If it is not for you, then you shouldn't do it. Mathematics is an enabler, not a bottleneck or gatekeeper. You have to be honest with yourself and get the necessary background.

When I was younger, we focused on a few areas and were guided by traditional mentors. Mathematics is much broader now and with many things going on between pure and applied mathematics. For example, my area of topology is now broadened to applied topology and topological statistics. Many classical areas of applications can be leapfrogged by other areas of mathematics that were historically theoretical. The excitement that gives and the potential impact on society is very important. The monastic view of a mathematician should be a thing of the past. Mathematicians should contribute to their subject, their students, and society. It is a challenge, but if you have a faculty position, then that should be treated as a rare privilege with an obligation to be a real contributor.

Young people should also understand that what you are doing now is not what you will do in ten years. If you are doing the same thing, then there is something wrong. There will be times when you will teach more, do more research, or do more administrative work, or even outreach or industrial connections. If you are open minded, then it can be enriching and give you more mathematical ideas and keep you intellectually awake.

AB: I always finish by asking about the future. Through your work at PIMS and now at Mitacs, you must be exposed to so many different trends in mathematics. What is on the horizon for mathematics and where do you think the subject is going?

AA: We are in a golden era of mathematics. Huge conjectures have been solved over the last twenty-five years such as Fermat's Last Theorem and the Poincaré conjecture. These are problems that were open for hundreds of years. The power of mathematics is evident, and its beauty is shining as bright as ever. We are attracting extraordinarily talented people into mathematics.

There is concern among young people about finding academic jobs and post-docs. Is the model of the university sustainable? Are we a technological breakthrough away from a substantial shrinking of the faculty base? We wrestle with these questions. You want to do what a robot cannot. Mathematicians know how to think and synthesize data, and they are open to connecting structures and seeing patterns and recognizing them. I think mathematics will play an important role in our society where data and thinking are combined.

Mathematics has a track record of its ideas being important after their discovery, even in the commercial world. It also possesses ideas that are extraordinarily beautiful. If we get the proper tools and are open minded, then mathematics is sure to be applied to all areas of knowledge. The use of quantitative methods in the natural sciences is becoming highly prevalent. The existing views we have in our head of what a mathematician is may change, and people will most likely be doing much more of it in the private sector than in universities. That may be a positive evolution, and we should welcome change.

Chapter 2

Interview with Federico Ardila

Federico Ardila is Associate Professor of Mathematics at San Francisco State University and Adjunct Professor at Universidad de los Andes, Bogotá, Colombia. His work focuses on combinatorics, with applications to algebra, geometry, topology, phylogenetics, and optimization. At the time of the interview, he was spending the semester at the Mathematical Sciences Research Institute as a Simons Research Professor. Federico is a Fellow of the American Mathematical Society and has held a National Science Foundation CAREER Award.

I first became aware of Federico through his 2016 *Notices of the American Mathematical Society* article "Todos Cuentan: Cultivating Diversity in Combinatorics", where he talks about his experience cultivating diversity in mathematics. We'll learn more about this and his community building between the US and Colombia in the interview. His "Course Agreement", which sets out ground rules for mutual respect and support in mathematics classes, is fantastic and something I've adopted into my courses.

This interview was conducted in February 2017.

$$\infty$$

Figure 1. Federico Ardila. (Photograph by May-Li Khoe.)

AB: What was your first mathematical memory?

FA: I remember some Math Olympiad tests arrived at my school in Bogotá, Colombia. I didn't really like math a lot in school, but I remember these tests had some innovative questions that I didn't know were math. I remember being excited about solving those. I was young, maybe eight or nine years old.

At the time, I think both my sister, Natalia, and my cousin, Ana María, had participated in these kinds of Math Olympiads; I was the curious little brother. I was lucky and privileged to have this material arrive in front of me at that age. It took me a long time to accept that I wanted to be a mathematician, but definitely the seeds were planted pretty early on.

AB: Was there a person or a teacher who'd influenced you to study mathematics as a career before you entered university?

FA: The Math Olympiads were a crucial experience for me. I ended up going to the summer camps and eventually making it to the Colombian national team. Later I coached the team for a number of years. I found a good community there: I had peers that loved this stuff, even if it was a small group. It was a rare chance to interact with kids from all over the country, and eventually from other countries.

Figure 2. Bogotá, Colombia. (Photo from Shutterstock.com.)

The person who founded and ran the Olympiad program was Mary Falk de Losada, an American woman who moved to Colombia in the 1970s. She had a big influence on me and many of the mathematicians of my generation in Colombia. She was also the first person I heard speak about discrimination in mathematics.

AB: How did you end up at MIT [Massachusetts Institute of Technology] for your Ph.D. and how did you come to work with your advisor, Richard Stanley?

FA: I arrived there as an undergraduate. I had never heard of MIT when I applied, and I didn't have the kind of grades that should allow me to get in. But a friend of mine wanted to go there and told me that they offered full scholarships. That's all I knew about it. I also remember their brochures had a kind of a punk rock aesthetic that intrigued me.

When I arrived in the US, the first class that I took was with Richard Stanley and with Hartley Rogers. They ran this seminar in preparation for the Putnam Competition, and I have to say that I had little concept of what a math major was. But I knew that I liked math competitions, and so I enrolled in them. I ended up being on the Putnam team of MIT, which they led. I was lucky to meet Stanley early on.

It's interesting because I hated combinatorics. I thought it was just hard. I always felt like I was good at the other stuff, but bad at combinatorics. I think there was a kind of stubbornness there; I wanted to master it.

As a senior, I took a couple of courses with Stanley and Sergey Fomin. And those two courses were just amazing, and they got me to understand that combinatorics was not just a bunch of clever tricks, but that there was a structure to it and a theory to it. That's when I fell in love with combinatorics in my senior year at MIT.

Figure 3. Richard Stanley. (Photo courtesy of the Archives of the Mathematisches Forschungsinstitut Oberwolfach.)

When I applied to graduate programs, I was still clueless. It didn't even occur to me to look at what a ranking of schools was. I

Figure 4. Sergey Fomin. (Photo licensed under Creative Commons Attribution-ShareAlike 4.0 International (CC BY-SA 4.0).)

asked Stanley for names of people I might work with. He mentioned four people, and I applied to those four schools. I was accepted to MIT, and everybody told me that that was the best place for algebraic combinatorics. I ended up staying there and working with Stanley. I also learned a lot from Gian-Carlo Rota, who was still alive at the time, and younger faculty like Sara Billey and Anne Schilling.

AB: Your research now is in combinatorics, with connections to topics like algebra, geometry, and even robotics. Would you tell us broadly what combinatorics is? What are some of the main goals of your research?

FA: I like the way my partner May-Li puts it; she's a designer who works in technology and education, and she loves mathematics. She calls combinatorics "the science of possibilities".

Many people say combinatorics is about counting. It's true that we do count things sometimes, but the way that I see it, we spend most of the time studying their structure. To give you a very simple example, we do not count the cells of a chessboard one by one: 1, 2,

..., 64. We first realize they have the structure of a grid, and then we use that structure to count them: $8 \times 8 = 64$.

Of course, I study more complicated objects, and most of my efforts are spent studying their inherent structure; sometimes those objects are discrete, and sometimes they're continuous. Once I've understood that structure well enough, then I can count them, or measure them, or prove other things about them.

As it happens in all of mathematics, there are some problems that are extremely hard but are isolated from the rest of mathematics. I always look for combinatorics that arises in or is inspired or motivated by other fields of mathematics. I think that makes for deeper and more connected mathematics.

Most of my work is related to Lie theory, to representation theory, or to algebraic geometry. I'm always trying to talk to lots of people in different fields to see how I can be of service as a combinatorialist. I like studying the combinatorial structure of other people's objects and seeing what new things we can find out.

AB: What research topics are you working on now?

FA: I'm spending the semester at MSRI, and it's such a luxury that many of my mathematical heroes and mathematical friends and students are living in Berkeley this semester and thinking about geometric combinatorics together. It's been a very productive semester, talking to lots of people, and sprinkling some seeds for future projects. The biggest challenge has been to balance many different projects.

One big project that I've been busy with lately, that I'm very excited about, is a project with June Huh at the Institute of Advanced Studies, and with Graham Denham at the University of Western Ontario. It seems we can prove a series of conjectures from the 1970s and 80s on the unimodality of some combinatorial sequences. In combinatorics, there are many sequences that are unimodal, meaning they start increasing, reach a peak, and then they come back down. When you study combinatorics, you're used to the fact that most of the sequences that we encounter have this shape. It's kind of an amusing thing to observe, but it's often incredibly hard to prove.

Some ways of proving these kinds of unimodality conjectures use algebraic geometry (using the hard Lefschetz theorem or the Hodge-Riemann relations) or representation theory (using the representations of the Lie algebra sl_2). What has happened often is that these kinds of conjectures are solved by taking these combinatorial objects that are easy to define, and revealing that, in fact, they have a much deeper algebraic or geometric structure.

We are proving some unimodality conjectures about matroids. Matroids are a combinatorial model of "independence", that unifies aspects of linear independence, algebraic independence, graph theory, matching theory, optimization, and Lie theory. I love matroids; they're one of these objects that are very connected to lots of fields of mathematics.

Matroids generalize linear independence in vector spaces. For example, the Fano matroid arises from the familiar Fano plane. June Huh, first in his Ph.D. thesis and later joined by his collaborators Eric Katz and Karim Adiprasito, was able to solve some very old unimodality conjectures for matroids using these kinds of techniques from algebraic geometry. For my taste, this was one of the most exciting recent developments in mathematics.

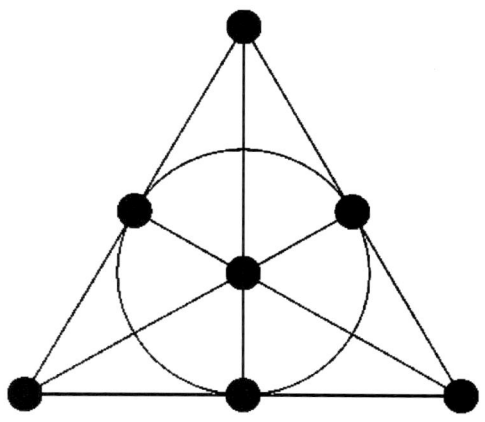

Figure 5. The Fano plane.

June, Graham, and I are further developing this Hodge theory of matroids to prove a stronger series of conjectures. Along the way, we have discovered a lot of interesting new structures, within and outside of matroid theory. We're at the writing stages, and everything is working out very beautifully.

AB: You're a professor at San Francisco State University, but you're also at Universidad de los Andes in Colombia. How does your work as a professor span the two institutions?

FA: Even though I had access to mathematics as a kid, I did feel foreign to mathematical culture, especially in the US. One thing that I always felt in US academic spaces is that they're very narrow in their conception of what's valued. We are advised that it's your theorems that you're going to be judged by, so they're the only thing that is important. I reject that, and I always have, and I remember thinking already when I was a freshman, "I'm going to have to do this my way because I don't believe in this idea of just being an academic, period."

I was raised to have an interest in the community, in equity, in trying to make a positive difference. Early on, I knew that I loved mathematics, but that I also wanted to have some bigger involvement with society. I don't want that to sound pretentious, but I just wanted to help a bit with whatever tools I had.

When it came time to decide where to place myself academically when I finished my Ph.D., I was intentional. Some of my mentors and peers disagreed with my choices. I wanted to be in a position where I could do high-quality research, and, at the same time, I could have an effect outside of research mathematics. I wanted to be in a place where I wanted to live, in a diverse city. I wasn't willing to move just anywhere because of academia. That's how I ended up at San Francisco State University [SFSU]. At the same time, I always wanted to do something with Colombia.

SFSU is a very interesting place research-wise for me. One thing that was also important to me and shaped my view of what an academic mathematician does is the community that it serves. More than half of our students are first-generation college students, more than half of them are first-generation Americans, and over seventy-five percent come from ethnic minority groups. When mathematicians visit me, they're often struck by our student population, because that's not what they're used to seeing on a US campus. But I don't think that's accurate; our campus is probably more similar to the average US campus than the typical R-1 university is. I think academic mathematics focuses on a very narrow view of who should be doing mathematics.

I love the Bay Area, and that is one important reason that I'm here. Also, I'm Latino and my partner is Asian, and few places have such a strong Latinx and Asian culture. When I found myself becoming comfortable in San Francisco, I decided that I wanted to set something up with Colombia so that I could stay in touch and give back to the community that gave a lot to me.

Figure 6. San Francisco and the Bay Bridge. (Photo from Shutterstock.com.)

I noticed that there wasn't anybody actively researching combinatorics in Colombia and there weren't courses being offered. I saw a need that maybe I could fill. I started offering joint graduate classes that were taught simultaneously at San Francisco State and in Colombia. This started as kind of a wild experiment; it was before all these online courses became a fad, and this was a very do-it-yourself setup.

I think it's a nice, productive bridge we created. The students not only took the course together, but they truly collaborated. For example, many of the final course projects were done internationally between a student in Colombia and a student in the US, and several became published research projects.

Eventually, I was awarded an NSF CAREER grant that funded my research and allowed me to make this a more systematic program. With that grant, I was able to keep offering these courses and to bring American students to Colombia and Colombian students to the US to do research together. It also allowed me to start the international Encuentro Colombiano de Combinatoria, which meets biannually.

I think that if you want to see minority mathematicians succeed, it's going to be easier to see ten of them succeed than to see just one. You can't just take one person, put them in a sea of otherness, and expect them to thrive easily if there's not a systemic change in the way things work. I wanted to create a wider environment where things were done differently, and there was a deep sense of community.

It's very exciting to me to see that my students in Colombia and the US are friends now, and they collaborate, and they're professors in many places, both in Colombia and in the US. It's become a wonderful community. In mathematics, we're measured by our theorems, and the mathematics needs to be on point. That's crucial. I'm very excited to see the mathematics that this community is doing. They're doing some beautiful stuff.

At the same time, I also like that they're not just doing mathematics and that there is this sense of activism in the community. We are finding ways to be involved with lots of different universities and even people in nearby countries. We are trying to create a different space where many people can thrive and do great mathematics.

AB: Last year you wrote an influential *Notices of the American Mathematical Society* article "Todos Cuentan" or "Everybody Counts". Would you tell us about the article and its message?

FA: Fundamentally, one thing that I believe very deeply is that mathematics is for everybody, and that's not how we have traditionally behaved as a mathematics society. Many of our practices are designed to select and support the "best" people. I try to take a different point of view. If a student shows up in my classroom, then there's a very good reason that they're there, and it's my job to support them.

Society has deep inequities. If we don't address those inequities very mindfully, then they're just going to be reflected in our classroom. That's one of the fundamental reasons why we struggle as a mathematics community to truly welcome diverse perspectives. We have to do our homework and learn about what are those inequities and what we can do about them. I think most math professors entered the university with absolutely no training on what it means to educate our society. We're trained on how to prove theorems, and we're often told that the educational part is not so important, and we shouldn't focus on it because it won't help us get the best job. I think I'm just trying to learn about pedagogy, and about the structural inequities that have taken us to where we are today.

A lot of it is about being thorough and scientific in the same way that we are about our science. What is the scholarship on really trying to make sure that we give equitable access to everybody? How do we move from just getting faces that look different in a classroom, to truly welcoming diverse perspectives?

There are many scholars that I'm very indebted to, who inspired a lot of what I wrote in that article; people like Audre Lorde, bell hooks, Paulo Freire, Estanislao Zuleta, Rochelle Gutiérrez, and Bob Moses. They have researched education and inequity more deeply than mathematicians have. I have also been blessed to be surrounded by wonderful people, doing work of a similar spirit in very different fields. My mother worked in violence prevention, my father in human management, my partner May-Li Khoe in design and education, my sister Natalia in music pedagogy, and my SFSU colleagues in science and in ethnic studies. My dear friends Sita Bhaumik work in art education and food activism and Dania Cabello in sports as a tool for social change. As in my mathematical research, I am always trying to learn from the practices of people in other disciplines.

One very important principle for me is that science is very powerful, and really shouldn't be concentrated in small sectors of the population. And I think science also brings a lot of joy and empowerment that should also be spread widely among our communities. I want to encourage mathematicians to constantly ask ourselves what we can do to make mathematics a tool towards a more equitable society.

AB: How did you converge to the agreement you include in your course outlines? What's the effect that it's had on students?

FA: I have always wanted to make the classroom a human place, where everyone is welcome, and where we don't only talk about mathematics. The obstacles to student learning are often not mathematical. I try to make my classroom feel like a very comfortable place, where people are welcomed to bring their full perspectives, and they feel safe taking risks and sharing their ideas.

One of my course outlines contains a version of the Agreement: "The goal of this course is to offer a meaningful, rigorous, and rewarding experience to every student; you will build that rich experience by devoting your strongest available effort to the class. You will be challenged and supported. Please be prepared to take an active, patient, and generous role in your learning and that of your classmates."

At the same time, I hope to make clear to my students that they're in the class not only for their own self but also to help and support others. This is good for the classroom, and it is important for their education. That's what's going to help them in society as they go and work in teams, and find out that they are valued not only for how they do but also for how they lift up the people around them.

I work to make sure that this doesn't feel like something I'm imposing on them; it is an agreement among all of us. We take the time for students to discuss what the agreement means to them, what they might add or improve on. I am always impressed by their openness and thoughtfulness in these discussions.

One central principle for me as an educator is to treat students with respect and communicate clearly and openly with them. Many of my better practices as an educator have come from really listening to them.

AB: What advice would you give to young people thinking about studying mathematics?

FA: One thing that I find very important is to recognize the joy of doing mathematics, to seek that joy constantly. At the same time, mathematics can be difficult and frustrating. And it's very important for a young person to know that it's not just them; even Field Medalists find math difficult and spend most of their time struggling. That's the nature of what we do. We're very curious, and we're never satisfied with what we already know. We're always looking for the next thing.

It's important to remember that the joy and the learning are yours. There's an expression that I like a lot in Spanish that I share with my students, "Nadie te quita lo bailado," and that translates to, "Nobody can take away what you've danced." To me, one thing this means is that no one can take away the joy with which you have done things. If you know that you love mathematics, then you can't let a professor or a low grade take that away from you. If you feel the joy and the power of understanding something new, then that joy and that power are yours, and they are real.

You are a good gauge of what you're learning, and sometimes the way that institutions measure you does not accurately reflect your potential as a mathematician. I can't deny that it is useful to learn how to test as well as you can. But I think it's also important to pursue knowledge with the purity of recognizing that you're learning something, that you're enjoying it, and that you're becoming a richer person for it.

AB: I always finish the interviews by looking forward. What would you say are some of the major directions in mathematics?

FA: I find it very hard to say. I know some mathematicians have very ambitious goals of solving big open problems. For me, I've more been driven by walking around the world, and seeing what I see, and trying

Figure 7. "Nobody can take away what you've danced."
Federico Ardila. (Photo from Shutterstock.com.)

to uncover something cool. If I see something interesting, I want to open that door and see what's there. Even though it is important to have these long-term goals for mathematics, I also believe that many of the most interesting developments and research directions didn't come that way. They came from unexpected places.

One thing that I think is important is disrespecting every border that people have tried to draw in mathematics. An ambitious student who is just starting out might try to find two fields that people think are unrelated and discover the relationship between them. Mathematics is interconnected in unexpected ways. I think the most interesting work comes from taking two islands in mathematics and showing how they're connected. That's a constant pursuit of mine. For me, it's centered around combinatorics, but I'm always trying to see how this field relates to something that it doesn't seem to be related to. I think that always leads to very interesting mathematics.

When it comes to the mathematical pursuit as a whole, one very big question I already mentioned is this: How do we make it possible for every community to participate in mathematics, benefit from mathematics, enjoy it, and use its power? This is a very important question.

In my mind, these two questions of mathematics and inclusion are related. Asking ourselves what are the most original mathematical developments of the future is closely related to asking ourselves who will make those developments. I want to see the most diverse group of people possible tackling the deepest mathematical questions. If you care about diversity and inclusion, you recognize that minoritized populations are often judged by higher standards, and you should tie your outreach efforts to the most interesting mathematics possible. If you care about the development of mathematics, you recognize that many of the most interesting discoveries have come from people who have not been indoctrinated into the ways that most mathematicians think, or who have dared to question those accepted ways of thinking.

My job gives me the opportunity to work with many students who have not been conditioned to think like most mathematicians. When they engage with deep mathematics, I often find that they think very differently from me, and ask very interesting questions that I hadn't asked myself.

I think it makes sense to think that mathematical research also works this way. New groups of people bring new perspectives and ways of thinking, and they might be the ones who see what everyone else has missed.

Chapter 3

Interview with Jennifer Chayes

I first met Jennifer Chayes at the 2012 Workshop on Algorithms and Models for the Web Graph conference in Halifax. She gave a keynote talk, and I chatted with her as she set up her presentation. My first impression was of her cool confidence and the force of her intellect.

Jennifer is one of the leading researchers in network science, working at the interface of mathematics, physics, computational science, and biology. She is the Managing Director of Microsoft Research New England and New York City. She is highly awarded, being a Fellow of the American Mathematical Society, a Fellow of the Association for Computing Machinery, recipient of the Anita Borg Institute Women of Vision Leadership Award, and winner of the Society for Industrial and Applied Mathematics John von Neumann Lecture Prize. She was also recently made Microsoft Technical Fellow.

This interview was conducted in September 2017.

∞

AB: What was your first mathematical memory as a child?

Figure 1. Jennifer Chayes. (Photo licensed under Creative Commons Attribution-ShareAlike 4.0 International (CC BY-SA 4.0).)

JC: I was four years old or so. I used to visit a neighbor—I started going there because a very nice woman gave cookies to my brothers and me. When we were choosing cookies, I heard her husband and daughter, who was a high-school math teacher, doing math problems. I thought it sounded cool, so I started asking them if they could give me puzzles. They probably thought I was weird, but they gave me puzzles, and I loved doing them.

There hadn't been any math in my household; my father was a pharmacist, and my mom couldn't add fractions (although she's very smart). But my neighbors sounded like they were having so much fun. They then started making up little word problems for me. I didn't know algebra or anything, although I did know how to count.

I liked it, and I found it very fun. They loved projective geometry, and they would give me things to work on at my level.

AB: Was there a person or teacher who influenced your early scientific career before university?

JC: In seventh grade, I took Euclidean geometry, and our teacher taught us how to prove things. It was an honors class for kids good in math. He taught us about logic: statements, converses, contrapositives, how to properly conclude things, and so on. I loved it, and for me it was magic.

This sometimes happens with great teachers. He obviously understood a lot—not all of my teachers were like that. In eighth grade, I asked my teacher why I couldn't put a square root in the denominator. My teacher said to me, "You can't put radicals in the denominator, just like you can't put bananas in the refrigerator!" While the year before I had someone teaching me formal logic. They were a huge variance in the quality, enthusiasm, and understanding that my math teachers had.

My seventh grade teacher was very passionate, and he would get excited when there were kids in the class like me who fell in love with the subject.

AB: Your undergraduate was in biology and physics, and then you completed a doctorate in mathematical physics at Princeton. What led you to Princeton, and how did you end up working with your supervisor?

JC: Theoretical physics was more stylish at the time than mathematical physics. People said mathematical physics wouldn't get me a job. But things have changed, with mathematicians now caring a lot about that topic, which is at the forefront of mathematics. I liked proving theorems, and I liked physics.

I thought I was going to be a particle physicist first, but it wasn't nearly as mathematical as it is now. Then I took a class from Elliott Lieb who was doing beautiful work on atomic physics. I liked the

Figure 2. Elliott Lieb. (Photo courtesy of the Archives of the Mathematisches Forschungsinstitut Oberwolfach.)

class, but I wasn't as wild about atomic physics. I asked him if he would do statistical physics with me as his student and he agreed.

I also worked with the supervisor of my ex-husband Lincoln Chayes, who was Michael Aizenman. I was co-advised by Elliott and Michael. The work with Elliott was more analytical, and the things with Michael were more probabilistic.

I knew Princeton had very good mathematical physics at the time. Even though I was not a mathematics undergraduate (I was one or two courses shy of such a degree), I loved mathematics. The summer before my senior year, I met Barry Simon at Princeton. Barry thought it was implausible that this couple coming from Wesleyan would go into mathematical physics at Princeton as they only accept one or two students in that area a year. He tried to have us apply to other graduate schools. I left feeling depressed, but I learned later that Barry was on the admissions committee and pushed to get us in.

AB: Your work is in many areas, ranging from graph limits, to phase transitions, to modeling complex networks. How would you describe your research to a non-mathematician?

JC: It might appear as if I am doing several different things, but I have a set of lenses through which I see the world. Many mathematicians have similar lenses, and, for many of us, the lenses are set early. In graduate school, I did work on random surfaces and percolation that helped set these lenses.

I try to see networks and geometric structures as representations of something going on in a certain system. Much of my early work was on percolation, which is like coffee percolating. A passageway is either open or closed to a liquid. Not everyone's grains of coffee are in the same configuration and yet there are some bulk properties similar to my espresso and yours. It's like taking draws from a distribution.

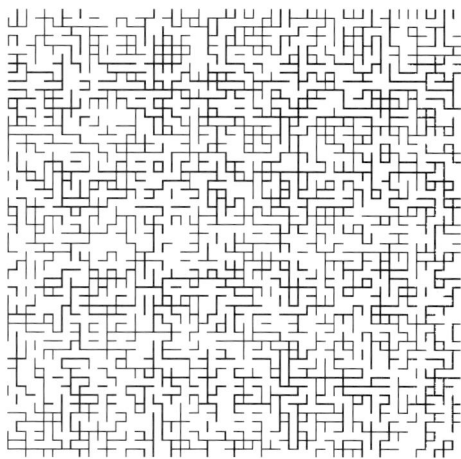

Figure 3. Percolation theory studies the connectivity of networks. (Image licensed under Creative Commons Attribution-ShareAlike 3.0 Unported (CC BY-SA 3.0).)

I tend to study systems with randomness, which is common in the real world. After I got to Microsoft about fifteen years ago, people were just starting to talk about the internet and the world-wide-web

as structures that one could understand. I took insights from percolation and phase transitions, and I used that to model the internet
and the web, and then later social networks. I also see networks in
computational biology that I do. There are omic networks, where
you have genomic or proteomic data, where many times you don't
observe the whole network. For example, when you study an evolu-
tionary tree, all you see are the leaves. In biological data, you don't
see every gene that has been activated or every protein in a cell. You
try to infer the missing parts of a network. These inference problems
are often related to machine learning problems. There are nice impli-
cations there: if I recognize a protein as important that no one else
has recognized, then that could be a drug target.

Phase transitions were things I also studied in graduate school,
and, for me, these have become a metaphor for the world. We just
underwent a phase transition in November 2016! In phase transitions,
when you vary the system, you see both quantitative and qualitative
changes. Examples are water boiling or freezing, or certain types of
magnets, which are magnetized or not at certain temperatures. They
also happen in graphical representations of other problems. This
is how I went from a mathematics department to Microsoft. There
were problems people studied in computer science that have graphical
representations and that undergo phase transitions from tractable to
intractable. There is a precise mathematical correspondence there,
and I started to use equilibrium statistical physics to study things
that were happening in theoretical computer science.

More recently, I've used non-equilibrium statistical physics methods, which happens in a system with a driving force of some sort. I
am using this to provide insight into how deep neural nets work.

AB: What research topics are you working on now? You can be more
technical here if you like.

JC: I am working on several different things. I am working on graph
limits, which is something we invented twelve years ago with László
Lovász, Christian Borgs, Vera Sós, and Katalin Vesztergombi. Christian and I have continued to work on the topic up to the present. We
first did graph limits for dense graphs, where every node is connected

to a positive fraction of all other nodes. Most networks in the real world are sparse, however. For example, Facebook keeps growing, but I'm not friends with a positive fraction of other nodes. So that's a sparse graph.

Figure 4. Jennifer Chayes with husband and collaborator Christian Borgs at Microsoft. (Photo licensed under Creative Commons Attribution 2.0 Generic (CC BY 2.0).)

In the last five years or so, we've developed two very different theories for graph limits of sparse graphs, and in particular sparse graphs with long tails like the Facebook graph or power-law graphs. We have one that is a static theory, a kind of L_p theory, where things are integrable, but they may not have a second moment. We also have a time-dependent theory that the statisticians like that models the progression of these networks.

Another thing I am working on is how to do A/B testing on networks. For example, we do A/B testing when the outcome of one group is getting a drug and another getting a placebo. Or the Microsoft homepage might roll out a different version to one percent of their traffic to see if they like that version more. But suppose I was studying people getting a flu shot with treated flu virus or a placebo flu shot. If your children got a real flu shot and you received a placebo shot, and none of you got the flu, it would not be sound to conclude the flu shot had no influence. There's interference in the network because members of your family are interacting with each other. So how do you do an A/B test on a network? We have methods to draw correct inferences. We are excited about it since it has many practical applications for the experimental design of tests on networks of interacting entities.

Another big project we are starting is with Stand Up To Cancer. They are a wonderful foundation that has raised about six hundred million dollars. For our project, they raised about fourteen million dollars. They bring together groups of researchers to study certain classes of questions. Usually, they bring together biologists and oncologists. Recently, they've started what they call convergent projects, where they bring mathematicians, physicists, and computer scientists together with oncologists and biologists.

Our project is called Convergence 2.0, and we are studying cancer immunotherapy. We are applying machine learning and network analysis to try to understand why certain people respond favorably to cancer immunotherapy while others don't. These are very complex problems involving genomes and your T-cell profile. Everyone has a different T-cell profile; there is a neat combinatorial trick your body does to come up with a unique T-cell profile. We will work with people at about ten different institutions on problems around this.

I also have a new physics-based theory of deep learning, which I mentioned earlier, although it's been non-rigorous up to this point. But in equilibrium statistical physics people are only now proving the results rigorously. There is little understanding of why these neural nets work, so I think it's worthwhile even to do non-rigorous work that gives us conjectures to attempt to prove.

AB: Congratulations on becoming a Microsoft Technical Fellow.

JC: In the industry, it's a big thing. It is the equivalent of a corporate vice president, but it is much more technical. Microsoft has over one hundred thousand employees, and under thirty Technical Fellows, so I am excited about it.

It wasn't just a promotion, but I have an additional responsibility. I have a lab in New York, one in Boston, and a small group in Israel, but I just got a new lab up in Montreal. It's a company that we acquired about eight months ago called Maluuba, which was roughly half research and half development, focusing on machine reading and comprehension, dialogue, reinforcement learning, and other topics in machine learning. I am also super excited about getting personally involved in the Montreal AI hub: Canada in general and Montreal, in particular, is at the forefront of AI, and I think this will continue. You have an amazing government, both federally and provincially in Quebec, which is supporting the Montreal region. We are going to be growing there, and I am thrilled to have a group there.

AB: Maybe someday you will start something in Toronto?

JC: Maybe. Like the investment in Montreal, the Canadian government is also making a big investment in Toronto with the Vector Institute. One of my post-docs is going there. There are three institutes in AI funded by the Canadian government. With these investments, instead of a brain drain, Canada is creating remarkable groups that can have an outsized influence on a dominant field.

AB: What advice would you give to young people, especially young women, on pursuing a career in mathematics and STEM?

JC: First, there is something I tell women even before they go to university: it's not sufficiently well publicized that STEM fields are creative and collaborative. We tend to see pictures of solitary guys sitting at computer terminals.

What I do is creative. I imagine worlds and prove theorems about them. Sometimes they have an impact on the real world; for example, they may help with cancer therapy. I have amazing collaborations, and I work in teams. And I have societal impact.

I think it is an easier life than being creative in other fields. I thought about becoming an artist when I was younger, but then I probably would have had to do a day job and come home exhausted trying to make art.

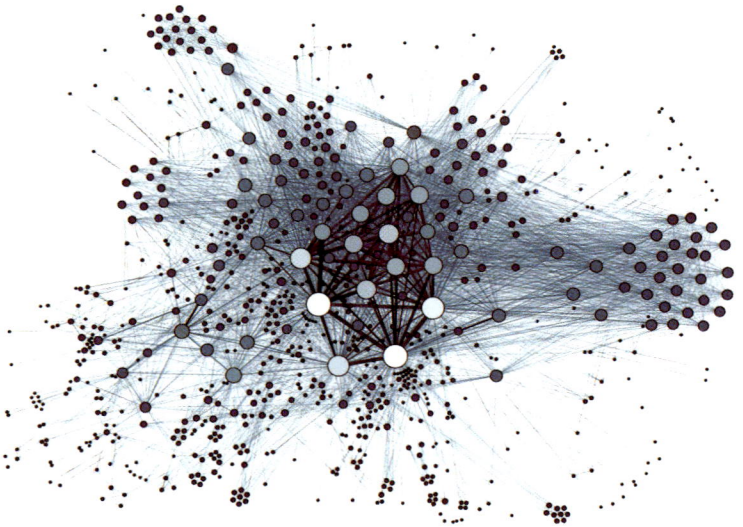

Figure 5. "I imagine worlds and prove theorems about them." Jennifer Chayes. (Image licensed under Creative Commons Attribution 3.0 Unported (CC BY 3.0).)

I would also say that for whatever reason, women tend to be less confident than men are. Part of that is not seeing as many role models. As a professor, I would interact with super talented women undergraduates, who didn't think they were good enough to go to graduate school. Women often don't realize that everyone is working hard; if you are doing well, it's also because of talent.

Women tend to take themselves out of the running before they should. Your self-assessment of your ability is not a reliable signal. If you like STEM, or if you had a teacher who tells you that you are good at it or you're passionate about it, then you should follow it. Reach out and network. STEM professions tend to lead to really satisfying careers.

AB: You were a child of Iranian immigrants to the US. What effects do you think the travel ban is doing to mathematical and scientific research in the US right now?

JC: I'm thrilled that Microsoft brings in Iranian interns, going through the necessary steps hiring them. We don't think about nationality when we are hiring someone. Many Iranians have been my interns and post-docs, so I was concerned by the travel ban. I have a young colleague at Yale whose family was stuck outside of the country owing to the ban. His wife and child couldn't get back in, and he couldn't leave because he had to teach.

I also have DACA [Deferred Action for Childhood Arrivals] students in New York City who are very scared now. I think we are a country of immigrants, and our greatest talent and vibrancy comes from that. Embracing immigrants is so fundamental to what it is to be America. I am concerned.

AB: I'd like to close with looking forward. What would you say are some of the major directions for mathematics in the future (or in your own program)?

JC: Whenever there are things that work well with little understanding of why, then I believe that there is some mathematics to be formulated and proved. In my position, I'm witnessing much of what is happening in deep learning—from image recognition to speech recognition to machine reading and comprehension. We see all these unexpected breakthroughs. These are high-dimensional random problems.

What is it about the structure of deep neural net algorithms that is finding useful information in these sparse high-dimensional structures? Answering that will involve many areas of mathematics, and a great deal of new mathematics will be developed.

I believe that the problems the world brings us guide us in the development of new mathematics.

Chapter 4

Interview with Maria Chudnovsky

Maria Chudnovsky is a leading mathematician working in the field of graph theory. She was born in St. Petersburg, Russia, in 1977, and moved to Israel with her family at the age of thirteen. Maria studied mathematics at Technion in Haifa. She completed her Ph.D. at Princeton University in 2003, supervised by Paul Seymour, where she is now a Professor.

Maria has the most famous Ph.D. dissertation in the recent history of graph theory. She proved, in joint work with Neil Robertson, Paul Seymour, and Robin Thomas, the "Strong Perfect Graph Theorem" (SPGT), which was first posed by Claude Berge back in the 1960s. Research on SPGT and perfect graphs resulted in hundreds of papers and partial solutions before its resolution. There are about 700 citations on MathSciNet for the search "perfect graphs" up to 2006 when the 178 page proof was published in the *Annals of Mathematics*. I vividly remember the excitement surrounding the announcement of the proof of SPGT, and how it sent ripples throughout the discrete mathematics community and beyond.

Maria is a giant in her field. For her work on SPGT, she won the Fulkerson Prize in 2009. She holds a MacArthur Foundation Fellowship (or "Genius" grant), and her research is funded by the National

Figure 1. Maria Chudnovsky. (Photo author: Gerd Fischer. Photo source: Archives of the Mathematisches Forschungsinstitut Oberwolfach.)

Science Foundation. Maria is also unique among mathematicians I know of for appearing in not one, but two television commercials: one for TurboTax and one for Comfortpedic.

This interview was conducted in May 2016.

∞

AB: How did you first become interested in mathematics?

MC: I don't remember a time when I wasn't. Math was always easy and fun, and everything else was hard. It was a very natural thing for me. It was always my favorite subject.

I remember the pain of learning to read, and I don't remember the pain of learning to count.

AB: Did anyone play a role in inspiring your interest in mathematics?

MC: My dad loved math. He was an engineer, but as a kid he loved mathematics. Probably he said enough things to get me interested.

I was lucky as I had very good teachers. I went to a special mathematics school in St. Petersburg, Russia. I was born in Russia, and my family moved to Israel when I was thirteen. I went to this school from the ages seven to thirteen, where math was the most important thing in the world, and the best thing you could be was to be good at math. I also had many good teachers who made things beautiful and made things interesting. From everything I heard in school, I never doubted there was anything more interesting than math.

AB: Can you tell us something about your experience at Technion? In particular, what was the environment like there, and how did it help lead you to Princeton?

MC: I started at Technion in the eleventh grade, where I began going to a Math Circle that was led by a Masters student in applied mathematics. It was a fun experience. He would solve Math Olympiad problems with us. If he attended an advanced mathematics lecture, he would tell us about it, making it digestible for us high school kids. It was also a social thing; it was where I met most of my friends. That was huge in my life. It was when I realized you could be a mathematician by profession.

Unlike in the United States, in Israel, you have to declare your major right away. You don't apply to a university, but, instead, you apply to a department. If I hadn't gone to this Math Circle, then I wouldn't have applied to the math department.

Both my parents were engineers, and it was always clear that I should go into engineering or computer science, as I was good in mathematics. This is ironic, as the math I do is on the border of computer science; many of my colleagues work in computer science departments.

I have to say, a lot of it was social. We were a group of friends, all very interested in math. We talked to each other about the things we learned, how pretty or nice something was. That reinforced my conviction that this is what I want to do in my life. If I could keep going like this, it would be great.

I applied only to three places for my doctorate. One didn't accept me.

AB: They are probably regretting that decision now!

MC: It was clear that Princeton was the right choice for me.

AB: Why did you choose to study graph theory?

MC: I knew I was going to study discrete math because as an undergraduate I thought it was a pretty area and I had a good intuition for it. When you like something, it is never clear if you first like it and become good at it, or you are first good at it and then like it. I certainly had this connection with discrete math.

At Technion, I also did a Masters in discrete math. All the places I applied to had strong discrete math programs. At Princeton, my research would have been graph theory, and, in other places, it would be other topics. I ended up at Princeton, so I became a graph theorist.

AB: How did you come to work on the Strong Perfect Graph Theorem (SPGT) in your doctoral work?

MC: I showed up at Princeton, and I knew I wanted to work with Paul Seymour. That was the problem he was interested in at the time. I was lucky at the time that I didn't understand what was appropriate and what wasn't. I came up to him and said: "Can I work with you on this?" He said "Sure."

Figure 2. Paul Seymour. (Photo courtesy of the Archives of the Mathematisches Forschungsinstitut Oberwolfach.)

When he saw I was contributing, I became part of the group. I don't know if when I first approached him, he thought it was a little bit strange.

AB: Andrew Wiles famously spoke about a great eureka moment when he settled Fermat's Last Theorem. Did you have a similar "aha" moment when settling SPGT with your collaborators?

MC: I was working with Paul Seymour at 5:45 pm (we usually work until 6:30 pm). We knew we were near the end of the proof, and there was one last step left. And then we saw it. We saw why A implies B.

We looked at each other and said "That's it. We are done; we can go home early."

AB: When you settled something as important as SPGT, were you confident the proof was correct?

MC: With all proofs, there are many levels of confidence. At first, you see a solution, and that is very good. But it is a huge proof so that you may have overlooked something. Then you sit down and write notes. After you have done that, you are more confident. And then you sit down and write a paper, and you feel even more confident. And then you start giving talks, and people think about what you said. And there is the refereeing process.

By now, we are pretty confident. With such a large proof, I don't know how to tell one hundred percent that it is true. Probably someone by now would have found a mistake, as it is something that people care about.

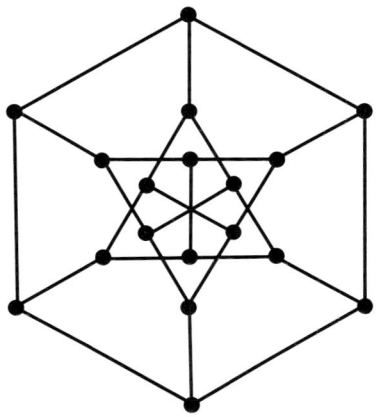

Figure 3. An example of a perfect graph. A graph is perfect if for every induced subgraph, the clique and chromatic numbers are equal.

AB: Structure plays a major role in the work you do, ranging from perfect graphs to claw-free graphs, or analysing other graph families. What do you think is the importance of structure in graph theory?

MC: First, to me personally, structure is very satisfying. You are not just answering one question; you are seeing some huge, global phenomenon going on. This is how I like the world to be, where I can completely understand what is going on. If I have a question, then I can go and look at this systematic picture that I have in mind, and I can find out the answer to this question. That is one kind of answer to your question.

Another answer is that it is surprising that some properties give you a kind of structure. There are many different properties of graphs you can think about. Some are little things: maybe they happen, or maybe they don't. Some have a huge influence on the graph.

In some sense, this is the strongest kind of theorem. You have some property, and it happens if and only if there is a certain structure. Somehow, that tells you a lot.

Not all properties allow you to prove a beautiful structure theorem. Some are just yes or no questions. When I give talks, I sometimes distinguish between properties and structure. It is remarkable and satisfying that with the property of a graph being perfect, you can understand the structure.

AB: What inspires your mathematical ideas?

MC: Things that appeal to me aesthetically. It could be a problem that seems beautiful or a concept that seems beautiful. Or it could be someone else's proof that seems beautiful, and I want to see what else I can do with it.

AB: The notion of beauty comes up often in our discipline. On that note, mathematics is sometimes called a science or art. What is your view?

MC: I think in the middle. Beauty is the guiding motivation. There is math that is motivated by physics, chemistry, or engineering. That is somehow separate. In much of math, you are just looking for the most beautiful thing you can think of. And only that determines if something is interesting or not. Beauty is also subjective. What I think of as beautiful someone else might think is ugly.

AB: What advice would you give to young people, especially young women, who want to study mathematics?

MC: I can give advice that is not exactly my own. When you start graduate school or anything big, you feel like that there is no way you are going to succeed. And there are setbacks: maybe you tried to solve a problem and you didn't, or you get a bad grade on an exam, or you attend a class you didn't understand.

You might then say to yourself that since I didn't understand this, I should be doing something else. That's not the right approach. What one shouldn't do is quit. It's not wrong to think about quitting, but I think one should take a very long time to consider the situation before quitting. Don't let your self-doubt scare you too much. Just accept that everyone has their moments when they feel like a complete misfit. Just keep pushing.

When you do something creative, ninety percent of the time you fail. If you are failing much of the time, you are not going to feel good about yourself much of the time. But then you succeed, and it more than makes up for it! You have to accept this as part of the creative lifestyle.

AB: Besides mathematics, what are your interests or hobbies?

MC: I like art. I don't produce it, but I like seeing art.

I have a two-and-a-half-year-old, and he is a full-time hobby. I am not one of these people who does math and then has another thing that is a close second. I would say my job is my hobby.

AB: Graph theory is a robust discipline with so many directions, such as structural graph theory, probabilistic graph theory, topological graph theory, and applications through network science. What do you think are the major directions in the field?

MC: That's a very good question that I wish I knew the answer to, as I would work in that direction. I think applications are becoming huge. I think applications are slightly different from the things I do since in applications the graph you are looking at is very large. The kind of things I do are deterministic things. What is needed in applications are more like if you assemble ten percent of your information about the graph, then what can you say with high probability?

Figure 4. Maria with husband Daniel Panner and son Rafael. (Photo courtesy of Emon Hassan/*The New York Times*/Redux.)

I think there are many people doing beautiful theoretical research that's vaguely or not vaguely motivated by that approach.

I would like to take the classical questions I've worked on and translate them into this language. We used to prove that *every* vertex of this set is adjacent to *every* vertex of another set. Instead, we can think about if *many* vertices of one set are adjacent to *many* vertices of another. I would look for an analogue or translation like that.

Yesterday I was on the train, and I saw someone with a t-shirt with a graph on it. And I thought, how nice. It was a Princeton Computer Science t-shirt!

Chapter 5

Interview with Fan Chung Graham

Fan Chung Graham (born in Taiwan in 1949) is one of the world's leading graph theorists and combinatorialists, with major contributions to spectral graph theory, random and quasi-random graphs, Ramsey theory, extremal graph theory, and complex networks.

Fan is a Distinguished Professor of Mathematics and of Computer Science and Engineering at the University of California, San Diego (UCSD), where she holds the Paul Erdős Chair in Combinatorics. She spent a sizeable portion of her career working in industry at Bell Laboratories and Bellcore. She published hundreds of papers and three books in mathematics and remains very active in research into her retirement. Fan won numerous awards for her mathematical work and was inducted as a Fellow of the American Mathematical Society in 2012.

I first met Fan in Vancouver in 2003 at the Workshop on Algorithms and Models for the Web Graph. My first memories of her were approachable, witty, and brilliant. Years later, I visited her home in La Jolla during a research visit at UCSD. I have great memories of walking through the cliff-top trails along the shore musing about the current and future directions of mathematics.

This interview was conducted in May 2016.

Figure 1. Fan Chung Graham. (Photo by Ché Graham.)

∞

AB: How did you first realize you wanted to study mathematics?

FCG: I loved geometry in middle school in Taiwan (which is equivalent to the 9th grade here). I was pretty good at puzzles. There was always this curiosity and problem-solving components to my work.

My father was the one who said when you are choosing your major, if you choose mathematics, then you can easily change to other areas, and not vice versa. He was completely right. Although I didn't change to other areas, the kind of mathematics I've been doing branches easily into subjects like computer science. Even now I hold a joint appointment in the departments of mathematics and computer science at UCSD. In fact, so many areas of computer science nowadays really involve beautiful mathematics, leading to good research directions. The situation is just like physics in the old days. A lot of inspiration now is coming from information theory and computer science.

AB: Segueing off that, can you describe how real-world applications such as complex networks have influenced your work?

FCG: I think combinatorics and graph theory, in particular, is a beautiful area but it also has connections with so many other disciplines. With the rise of network science, graph theory plays a vital role in laying the foundation for the rigorous analysis of modelling real-world networks. In the other direction, we feel the influence also: problems in information networks enrich the research in graph theory. They often point to many central directions.

For example, random graphs can be used to study real-world networks. Classical random graph models use equal probability distribution on nodes and edges; each has equal importance. This doesn't happen in real networks. This idea opens up a whole range of problems and directions in studying extremal and random graph problems in a much richer context.

AB: What led you to work with Herbert Wilf as your doctoral supervisor?

FCG: Herb had a huge influence on my research and career. I was lucky to run into him. I had never really taken classes in graph theory before I met Herb. Before I met him, I didn't even know of the existence of such an area.

I remember the day vividly when he walked into the office of graduate students where I was working. Later on, he told me he usually approached doctoral students who passed their qualifying examinations with the highest scores. At the University of Pennsylvania, I was a very good student, so he came to talk to me.

Figure 2. Herbert Wilf (1931 - 2012). (Photo licensed under Creative Commons Attribution-ShareAlike 3.0 Unported (CC BY-SA 3.0).)

He introduced me to combinatorics by telling me research problems. I had a good background in mathematics when I was an undergraduate at National Taiwan University, where I took many advanced courses. With Herb, I started to learn how to do research. I used what I learned to attack research problems.

He gave me many good problems. In our first meeting, it was a Ramsey number problem, and that topic is always delightful. It was a multi-color Ramsey number problem, which was an old one. At that time, I didn't know its history. I just picked up the problem and found a better bound.

AB: How did you do that? Did the ideas for the proof just come to you?

FCG: You know how it works with mathematics: you just work with pen and paper. I was just playing around. Ramsey problems are like a puzzle; in combinatorics, it is about finding or avoiding patterns. These are classical problems.

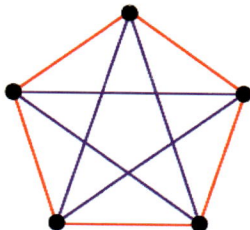 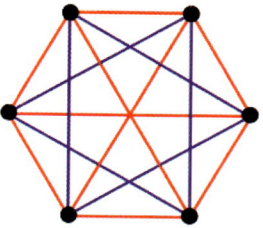

Figure 3. In Ramsey theory, we find unavoidable patterns, such as monochromatic triangles in edge-colored graphs.

My result is still the best one to date for triangles with multi-colors. At that time, I didn't realize that. I just solved a problem! Herb was very enthusiastic hearing my solution and encouraged me onwards. After that, I started working with Herb.

Herb had a gift of choosing terrific problems. He was an excellent teacher but also a great mathematician. Recently, I co-authored an article in the *Notices of the American Mathematical Society* about him. As a student, I solved one problem of his after another; altogether, I solved four problems. It was tremendously enjoyable working with him.

AB: You've had an impact on many areas such as spectral graph theory, Ramsey theory, and complex networks. Did you move between these areas by accident or design (or both)?

FCG: Both. After I received my Ph.D., I started working at Bell Labs. At that time, we had a lot of interesting people coming through, like Paul Erdős and Joel Spencer. I was doing extremal graph theory at the beginning, and it was much like the Hungarian school; they are so strong in that area.

In looking back, I moved from one area gradually to another. One area that I think is quite central is quasi-randomness. In extremal graph theory, you usually study one graph property and relate it to another. For example, if a graph has too many edges, then some pattern appears, like cliques or complete bipartite graphs. At some point, it was quite natural to take all these properties and put them in their rightful places.

The study of quasi-random graphs is used to identify these graph properties that look different but are in fact equivalent. These properties happen to be satisfied by random graphs. If you want to validate one property, then you choose the easiest one, and you get them all. Quasi-random graph properties unified random and extremal ideas. To me, randomness in graphs is defined by properties of graphs.

We worked on quasi-random graphs, hypergraphs, and other structures like sequences. But although we wrote all these papers (on quasi-randomness), not many people were following up (at the time), although I know it was a really good topic. Maybe we wrote too many papers in this area. We decided to stop for a while. Ten years later, there is a huge following in the area, with results cited in at least four textbooks, and a lot of new developments.

Then I moved to spectral graph theory, which I think is a major graph invariant. Think of a complicated structure far away, like a distant galaxy. We want to use relatively few parameters or quantities to nail down its properties. Spectral graph theory uses eigenvalue or the spectral gap to tell the shape of the network.

From the very beginning, I realized spectral graph theory was central. But the classical work in the area was algebraic, using tools from group theory and linear algebra. If the graph is very symmetric (for example, vertex- or edge-transitive, or even distance transitive), then the spectrum is short with high multiplicity.

Nowadays, in real-world networks, you can't see the whole structure, and these networks are far from regular. For example, think of internet networks or social networks. I realized that spectral graph theory could be viewed geometrically rather than algebraically. The algebraic approach has clear rules, and you nail down things precisely. Geometrically, you want to simplify the structure... just get the first order of things. In particular, I realized the connection with spectral geometry, its continuous counterpart.

The spectral graph theory I am doing is great for working with general graphs; that is, non-regular ones. The spectral gap will capture its shape, rather than in the usual adjacency matrix which might capture a few large degrees. To use eigenvalues to study the shape of the network, we had to use the normalized Laplacian. The formulation is a bit complicated; it isn't just a 0-1 matrix. But the normalized Laplacian relates to random walks, diffusion, and the shape of the network. I knew it was harder to get this formulation accepted by other graph theorists, but I realized it was the right way to see it.

In spite of the early resistance, it was clear to me that the normalized Laplacian was the right way to proceed. In the beginning, most people used it in computer vision and optimization, as they had to deal with subgraphs. Subgraphs of regular graphs aren't usually regular.

Now, my approach to spectral graph theory has been accepted, especially in the algorithmic community and in complex networks. If the eigenvalues are known, then you can make a lot of predictions about a network's structure.

Looking back, I moved from one area to another, but the underlying mathematics is connected, and not so different.

AB: You worked in academia and in industry at Bell Laboratories and Bellcore. How would you compare working in industry versus academia? Do you prefer one to the other?

FCG: I had a great experience at Bell Labs. In the old days, we had so many good people. Those good days, however, are in the past. It was a very different environment compared to a university. Imagine having 3,000 Ph.D.'s working under one principal investigator all under one roof. In those days, we could recruit the top people, beating even top universities like MIT or other places.

Because of my research, I made the transition to academia quite easily. Ingrid Daubechies and I were some of the first women to get tenure at Ivy League schools; I moved to the University of Pennsylvania, and Ingrid Daubechies went to Princeton. The academic world is thoroughly enjoyable with great students. But it is different.

In Bell Labs, we tackled big projects sometimes. That function is now taken over to some extent by start-up companies. In the mathematics center at Bell Labs, we usually had smaller projects like our research papers on topics of our own interest.

Another great thing about Bell Labs was that you could easily go across the boundary of one discipline to another: problem-solving has no boundaries. You need to use tools from all directions. My office was next to Michael Garey and David Johnson who worked in computer science, and I moved in that direction quite easily.

AB: Cross-disciplinary work happens quite a bit in academia, but it does feel that departments are too cloistered.

FCG: Not only because of the department division lines but also because of the peer-review process. You need to define which area you are. That is not helping you break into other areas. But still, peer review is the best way. For example, democracy is not always efficient. But what is the alternative? At the big labs, those division lines were not there. At UCSD, most of my co-authors are in computer science, which is a great source of great problems, especially in combinatorics.

AB: You and Ron Graham were some of the best-known collaborators and supporters of Paul Erdős. What was Erdős's impact on your work and life?

FCG: That is a very good question. I look back and reflect on the papers from when I worked with Paul. In extremal graph theory, about half of my work was directly or indirectly with Paul. I have joint papers with him, but mainly I got problems from him. He was like a bird picking up seeds from different places. Whenever he came through, he would mention problems he heard from other places. If Paul was interested in a problem, then there was extra motivation to work on them.

One example is my work on graph pebbling, which is a popular problem now. I first heard that problem from Paul; it wasn't his problem, but he told me about it. I liked hypercubes, so I worked on it and wrote a paper. I included a conjecture at the end, which is now called the Graham conjecture. In fact, I had a casual conversation about it with Ron, and I put his name on it! Usually, at the end of the paper, I like to put conjectures.

Figure 4. Paul Erdős (1913–1996) with Ron Graham and Fan, 1986. (Photo licensed under Creative Commons Attribution 3.0 Unported (CC BY 3.0).)

That summer Joe Galian ran his Research Experience for Undergraduates (REU), and he asked for some of my papers. I sent him the pebbling paper, and that did the job. The students all started work on it, and other REUs at other places did too. Now, there is a huge following for this area.

AB: It was the seed that Erdős brought that you germinated, and it grew into a garden.

FCG: Exactly. I heard this problem from Erdős, gave the problem to Galian, and now every time I go to conferences there are papers in this area. This is one example of how Erdős's problems propagated.

AB: What advice would you give students studying mathematics, especially young women studying in mathematics?

Figure 5. Joseph Galian. (Wikimedia Commons, public domain.)

FCG: My answer is simple: don't be intimidated. In mathematics, you can build up one step at a time. Once you do it, it's yours. It is a big area, and no one knows everything.

AB: I am familiar with your painting and musicianship (I've seen you play piano and Chinese harp or guzheng). What draws you to the creative arts?

FCG: It works in a different part of your brain than mathematics. It is nice to transition between the two. I enjoy watercolors and guzheng. I think it helps me; it complements rigorous mathematics thinking.

Mathematics and art are related in different ways. In watercolors, the work is done in layers. You have to wait until it dries. The color and texture can be different from what you expect. The challenge is to partition your picture into different layers, adding a conceptual viewpoint of how they all fit together. Painting is an interesting, unpredictable art form.

I started by doing landscapes and seascapes. I didn't get too much progress, so I spoke to my painting teacher Maria Klawe, and also started to do portraits. I did portraits of mathematicians who influenced my work. Portraits are much harder as our eyes are so critical of human faces. I added one layer after another, and a stranger

Figure 6. The guzheng is a traditional Chinese instrument. (Photo from Shutterstock.com.)

stared at me back. A person emerged that was different than I intended, so I had to start again. It's very interesting to reveal some aspect of your painting subject. You can't include every aspect, but if you grab some aspect of the person, then it is a success.

It's like mathematics: when you are proving theorems, you solve a small piece of the puzzle. Sometimes after you have done enough, there is a chance of developing a theory.

AB: What are some of the major directions now in mathematics?

FCG: In the old days, physics was a major driving force in mathematics. Nowadays, the new directions are often coming from data and network science. Combinatorics is serving as a bridge to these areas: bringing mathematics to them, but also bringing problems to mathematics. That influence is going to be a dominating force and will be wonderful for mathematics.

Figure 7. Fan's portrait of Paul Erdős. (Photo courtesy of Fan Chung Graham.)

Chapter 6

Interview with Ingrid Daubechies

Ingrid Daubechies is the James B. Duke Professor of Mathematics and Electrical and Computer Engineering at Duke University. Her doctoral work was in physics, focusing on the mathematics of quantum mechanics. She is best known for her discovery of compactly supported continuous wavelets, now called *Daubechies wavelets*. Her work has found broad application to signal and image processing, and more recently to art restoration and detecting art forgeries.

Ingrid has a luminous academic career, and she has many accomplishments and awards. She won the Leroy P. Steele Prize for Seminal Contributions to Research, the Ruth Lyttle Satter Prize, and the Leroy P. Steele Prize for Mathematical Exposition. She was the first woman to receive the National Academy of Sciences Award in Mathematics, and she's a former MacArthur Foundation Fellow, Guggenheim Fellow, and member of both the National Academy of Sciences and the National Academy of Engineering. She is a Fellow of the American Mathematical Society and past president of the International Mathematical Union from 2011 to 2014. Most recently, Ingrid was awarded a major grant from the Simons Foundation for US$1.5 million. Recently on December 7, 2016, Ingrid received the

Figure 1. Ingrid Daubechies. (Photo courtesy of Duke Photography: Less Todd.)

Gold Medal of the Flemish Royal Academy of Arts and Sciences for her merits for Flanders, the Dutch-speaking part of Belgium.

Ingrid gave truly memorable answers to my questions. My favorite part of her interview was when she turned the tables on me and asked me if I've met many women plumbers. In my view, she is one of the most impactful working mathematicians, and among the most prominent women mathematicians alive today.

This interview was conducted in December 2016.

∞

AB: When did you first realize you had a talent for mathematics?

ID: It didn't come suddenly. I was always interested in mathematics, and how and why things work and fit together.

AB: You also had an early interest in physics, and that was the subject of your doctorate.

ID: Yes. I don't have a degree in mathematics.

AB: Was there a person or teacher who influenced your mathematical development?

ID: My father took great pains to answer my scientific and mathematical questions if he knew the answer. I had a physics teacher in high school who I very much liked, who went into explaining. Most people would find that the teachers who influenced them were those who went beyond the textbook or recipes. The best teachers are those who convey understanding and insight and stimulate questions. I had a mathematics teacher in high school who was also like that.

At university, I considered switching my major from physics to mathematics. But then I took a course on Fourier optics that was absolutely beautiful. If you look at what a lens system does mathematically, then it turns out that it computes a Fourier transform, but in an analogue way. I thought that was beautiful. That made me decide to stay in physics. The professor for that course was Roger Van Geen; he passed away years ago, but he was a very distinguished engineering professor (he was a physicist) at my alma mater. All my training, both undergraduate and graduate, was done in Belgium. I came to the US for the first time as a post-doc.

AB: Was it your interest in physics and mathematics that led you to a doctorate?

ID: In Belgium, it was the case that if you go for a Ph.D., then you often do that at the university where you did your undergraduate degree. Professors are on the lookout for smart students, and if they have funding with which they can support a graduate student, then they contact students to ask if they are interested in a Ph.D. That is what happened to me.

Coming to the US, I've been exposed to the idea that it's a good thing to change schools between your undergraduate to your graduate degrees. I do think now that is good advice. When I tell students in Belgium that, however, they say that they can still learn so much from their professors, and they ask, "Why should we change?" I respond that apart from learning material, there is also the learning of an outlook, or way of approaching things, that you subliminally pick up from your professors. It is good to learn new approaches from other people. They don't believe me, by and large.

AB: You've worked in both industry at AT&T Bell Labs, and in academic settings at universities such as Princeton and now Duke. How would you compare the culture of both settings?

ID: I would preface my answer by saying that AT&T Bell Labs was the closest environment to an academic group outside academia, and the furthest from a consumer product oriented industry lab that I know. AT&T Bell Labs was big then, with about ninety percent of work focused on development, and ten percent on research. I was on the research side.

Paradoxically, I had more time for my research at Bell Labs than I would at a university: for example, I had no teaching or committee work. As time went on in my career, I have assumed more responsibilities in my profession nationally and internationally. That has left me less time.

On the other hand, I like teaching, so I missed the contact with students at the labs. I did miss contact with academics and intellectuals other than scientists or engineers. There were fantastic engineers and scientists at Bell Labs, but I thought it would be nice to meet art historians, political scientists, and philosophers. I realized that progress in my career meant that when I returned to academia I had less time than before, and it took as much as ten years before I had the breathing space where I could find the time to talk to non-STEM academics.

Every university has its own atmosphere. Princeton is a fantastic place and one of the oldest universities in the US, while Duke is a younger institution. At Duke, I experience the administration as much more attuned to the faculty. Princeton is a place directed by an inner circle. For example, at Duke, there are more creative approaches to offers to incoming faculty from other institutions.

Figure 2. Duke University. (Photo from Shutterstock.com.)

AB: One of your most famous discoveries concerns wavelets. What is a wavelet? How do wavelets apply to image and signal processing?

ID: Many things of interest to mathematicians or engineers are very complex and have several complicated structures layered on top of one another. As we do in other things in life, we like to take things apart into more elementary building blocks so we can unravel this rich structure.

To do such unravelling, we view these complex things as a super-position of much simpler ones. You could imagine trying to work on a screen or tablet with graphical tools putting down broad strokes of color, and then gradually adding finer details. What you are doing is constructing a complex drawing with elementary tools or action.

Decomposing into simple building blocks can be done in many different ways in mathematics. Wavelets are one way that emerged gradually as a powerful tool in this direction both theoretically and computationally. It is powerful because it can capture many different scales. For example, in an image or photograph, there are large color planes and small details such as edges or textures. For large color

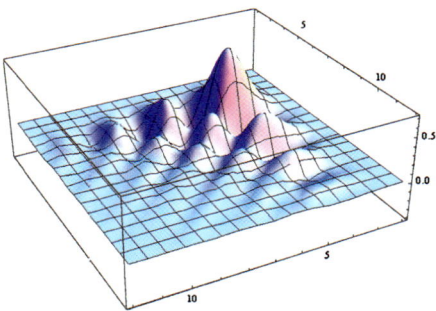

Figure 3. A Daubechies wavelet. (Photo from JonMcLoone at English Wikipedia. Licensed under the Creative Commons Attribution 3.0 Unported (CC BY 3.0).)

planes, you might want to use a building block that gives that feature. As you go to smaller and smaller detail, you try to take more localized ways of making changes to your image. Wavelets are mathematical building blocks that allow you to decompose complicated things in such a multiscale fashion.

Wavelets are computationally very useful, as you can start with your image and think of the fine details as all the information you lose when you blur the image. All that information is fine and local. You can do that again and blur even more, and so on. It is like peeling off layers, and each layer has its building blocks. It gives you a fast way algorithmically of making a decomposition.

AB: Wavelets also have applications in JPEG files?

ID: Older JPEG files use a discrete cosine transform, but JPEG 2000 (used for many current internet applications) uses wavelets.

AB: Your work spans both mathematics and its applications to science and engineering and beyond. What do you think the role of mathematics is in the sciences and engineering?

ID: Mathematics is what you need to quantitatively and accurately describe phenomena in science. Mathematics is the discipline we invented to be able to do that. When people describe the "unreasonable effectiveness of mathematics", they seem to say mathematics is more effective than other things. But we have nothing else to do that. I am not saying there aren't other things in life. There are tons of things in life in which mathematics plays no role. But to accurately and quantitatively describe things in science and engineering, mathematics is what you need.

We have built many mathematical frameworks as we realized that certain ways of understanding things are useful in more than one context. That is when you start abstracting, giving names to mental concepts in one field or another. That is how you build mathematics.

At some point, you start enjoying the mathematics—just the building of these mental concepts. That is what pure mathematicians do: they work only with those abstract concepts. It is similar to when artists play with materials and build things that others enjoy.

AB: Congratulations on receiving the Math+X award from the Simons Foundation, which comes with a US$1.5 million grant. Can you tell us how you came to receive this award and how you plan to use it?

ID: The Simons Foundation has, for a number of years, given these awards. They ask universities to nominate someone, and the nominee has to write a five-page description of how they see themselves, their research, and what they want to do. I am fascinated by applications to engineering and other fields; I am also fascinated by current developments in machine learning. Although I am in awe of many of the achievements of machine learning (such as recognizing faces), I think there is a challenge in understanding why these machine learning algorithms work so well.

Right now in machine learning, if you give the algorithm lots of data that is labeled (for example, representing one species or another), it will churn through these and learn how to distinguish type A from type B. In the end, you have something that is very complicated, but we haven't gained much insight into why A is different from B. In other situations, we don't have lots of type A or B data. I would like to understand machine learning at a higher level and use that in our scientific investigations. I described some of those in my proposal. I work with biologists, art historians, and so on, and I described those collaborations, and they liked it.

AB: In my recent interview with Nassif Ghoussoub, he mentioned that more mathematicians need to be working on machine learning.

ID: Some mathematicians feel that if something is too complicated with no nice description, then you cannot do good mathematics with it. However, I feel that if something works and achieves results and we don't understand it, then we need to work on it. I agree that we need more mathematical work on machine learning.

AB: Would you explain your work and interest in using mathematics to restore art and detect forgeries?

ID: I had a colleague who was interested in art, and who thought that image analysis could be used for art history and conservation. He convinced people interested in image analysis and art historians to create joint workshops, which were quite interesting. The first application was to detect forgeries; although we had some good results, that was not the most interesting of the projects I've worked on in this direction. Once they saw what we could do, others would approach me with other projects.

For example, some projects were trying to get a better impression of a painting hidden under another painting or removing cracks virtually from a painting to get a better reading of the work. The projects were mathematically and engineering-wise quite challenging, and different from other fields. These provided excellent projects for my students and provided interesting results for art historians and conservators. We also use machine learning for these projects.

Figure 4. Ingrid lecturing on her work using image analysis in art restoration. (Photo courtesy of Ingrid Daubechies.)

AB: What role does beauty play in mathematics? Do you have views on mathematics as an art or science (or neither/both)?

ID: Think about what it means to find something beautiful. For me, I recognize in myself a feeling of joy, awe, and reverence. It gladdens the heart and lights up the eye. Mathematics is beautiful as it elicits the same emotions in us. We may recognize it as elegant or different. It is awe-inspiring and joy-inspiring at the same time. For example, consider the experience you have after understanding something arduous, glimpsing the truth, or a way of doing things in a new way; with all these things you get that feeling.

An element of surprise is also important. Beauty always has a surprise. I don't find it strange to talk about mathematics as beautiful. It's not just mathematics; biologists come up with extraordinarily elegant ways of testing a theory or devising an experiment. Economists when checking a theory have to come up with ways to test a hypothesis, and some of these are quite elegant. Beauty comes up in many abstract disciplines.

To me, the fine arts (music, painting, and so on) are a shortcut to part of the human emotional experience. It gives us access to part of ourselves on our emotional plane. Mathematics is experienced in the intellectual plane and is a tool to do things intellectually. In that sense, mathematics and art are different, but this notion of finding beauty in mathematics is on the emotional plane.

Saying whether mathematics is an art or science is a way of trying to capture something with a sound bite. Using only one word to describe it simplifies things too much.

AB: The state of women in mathematics and STEM is of active interest these days, with many mathematics departments not even reaching twenty percent of their tenured faculty as women. What advice would you give to young women studying mathematics? Also, what advice would you give to departments to help achieve gender parity?

ID: Most people agree that women are just as good at doing mathematics as men. If you look at the papers that are written, it would be hard to decide from the math whether a man or woman writes it. The number of women in mathematics depends enormously on what country you consider. Even in Europe, neighboring countries can have completely different ratios of men and women mathematicians. I think this is more of a cultural thing than anything else.

If you are a young woman, you should not believe anyone who tells you that mathematics is not a job for women. To become a professional mathematician, you have to love mathematics. As an academic mathematician, the thing you will get the most recognition for is your research, which can be a challenging thing. You spend ages trying to figure things out. When you figure it out, you are overjoyed, but then you realize you want to understand it better. As you understand it better, you feel like a moron for not having seen it sooner. Research has fantastic highs that are not very frequent, and they don't last a long time. If you love mathematics, then that is good enough for you. You can be a good mathematician as a man or woman.

Today we had a funny experience, where the faucet was broken down on the floor of our department, and we had a plumber come and repair it. I've never thought about it, but it was the first time in my life I met a female plumber. Have you seen many female plumbers in your life?

AB: Not too many.

ID: I remarked about it to her, and she said she hears it all the time.

AB: Maybe you were the first female mathematician she's met.

ID: She didn't say that. There are many other pursuits in life that woman can excel in. We are accustomed to certain cultural things.

I remember a joke from a while ago when Margaret Thatcher was prime minister in England. A kid on television was asked what he wanted to do. He said maybe, fireman or astronaut. They asked him "Not prime minister?" and the boy responded, "Nah, that's a woman's job." I hope these cultural shifts are persistent because we react to what we see.

AB: I just wrote about Emmy Noether getting a statue. And you were involved this year in unveiling a plaquette of her for the ICM [International Congress of Mathematicians] Emmy Noether Lecturer.

ID: Yes, and the likeness of her on the plaquette is very good. And there is a bust of her in Munich. It's not very representational of her, however.

AB: I'd like to close with looking forward. Given your broad interests, what would you say are some of the major directions for mathematics and its applications in the future?

ID: I don't like to make broad predictions. For me, machine learning is my last big project. On the other hand, when I was at the IMU, I was trying to help developing countries foster a more mathematical culture. I think it will help them with their economic development,

Figure 5. The Emmy Noether Lecturer plaquette, designed and sculpted by Stephanie Magdziak. (Photo courtesy of Stephanie Magdziak.)

as it will train better engineers and scientists. I think having good mathematicians helps the economy. There are economists that have studied the impact of having good mathematics training, and who feel there is a definite connection. I hope the mathematics community will help that effort in the coming decades.

Mathematics is also at a very exciting time, with so many cross connections happening between areas of mathematics and other fields. It is a very fun and fertile time to be a mathematician.

Chapter 7

Interview with Nassif Ghoussoub

Nassif Ghoussoub is the founder and current director of the Banff International Research Station (BIRS), the founding director of the Pacific Institute for the Mathematical Sciences (PIMS), and the co-founder of the Mitacs Network of Centres of Excellence (NCE). On top of all that, he is an award-winning mathematician, whose most recent work focuses on differential equations and mass transport theory.

I met Nassif when he was the Scientific Director of MPrime NCE. My first impression of him was that he was amiable, had a laser-sharp mind, and was a natural leader. I've read his blog *Piece of Mind* for years, where he posts refreshingly blunt but fair views on the academy. Nassif was recently inducted into the Order of Canada, which is an honor he very highly deserves. I think we should all be proud as Canadians to have a mathematician of his stature represent us in Ottawa and on the world stage.

During the interview, Nassif was warm and open. I found the way Nassif talks about mathematics and the creation of PIMS and BIRS truly inspiring.

This interview was conducted in October 2016.

∞

Figure 1. Nassif Ghoussoub. (Photo courtesy of Nassif Ghoussoub.)

AB: Congratulations on recently receiving the Order of Canada. How did that transpire, and what was the experience like entering the Order?

NG: A few weeks before Christmas of last year, I received a phone call from the Governor General's office. I was asked if I would accept to be an Officer of the Order of Canada if appointed by the GG. First, I thought it was a prank. How can one refuse such an honor? It was a big surprise. I was ecstatic. For me, it is a huge deal, mostly because of the way I came to this country. I was essentially a refugee, though I didn't enter Canada as a refugee. Two months before I was about to defend my doctoral thesis in Paris, a civil war broke out in Lebanon—my home country, where I was planning to go back, teach at the university, and be with my family. I came instead to the US as a post-doc, to wait the civil war out, then to Vancouver, where UBC

[University of British Columbia] had offered me a position. The war didn't end until sixteen years later. By that time, I was a fully-fledged Canadian with a Canadian family. I feel blessed.

Besides giving me refuge, Canada allowed me to contribute. My friend Ivar Ekeland (a former president of Université Paris-Dauphine, who eventually came to be director of PIMS) told me once that I couldn't have done in France what I have managed to contribute here. He has lived both the Canadian and French contexts, and he is aware how closed and rigid the system is in France, while Canada has so much potential to move forward. All that we need here is more people able and willing to participate in building this great country. The opportunities are incredible. Just look at the Trudeau government with its young cabinet ministers, some of whom are relatively recent immigrants.

I've had accolades in the past, but, for me, the Order of Canada was really special. This is the country where I lived the bulk of my life. I developed so many friendships over my forty years here. Another enjoyable byproduct of this honor is how it triggers a flood of congratulatory messages from so many of them, reminding you of the many awesome people you've known at various stages of your life.

AB: What was your path to becoming a mathematician?

NG: I read your interview with Ken Ono, and it was striking to me how different our paths to mathematics were. We do have a somewhat similar refugee background, but Ono was born into a mathematical family and was immersed in the mathematical culture. His father was a serious mathematician. I come from a small mountain town of mostly farmers (Beit-Chebab) in the Levant. My parents didn't have a chance to even finish high school. They married very young and had to leave right after World War II to western Africa to make a living. Very early on, however, they wanted us to get an education, so they sent me back to Lebanon with my siblings to Christian boarding schools, when I was only three years old. I didn't have the cultural or family background even to know what mathematics is all about (as opposed to arithmetic), never mind advanced research in mathematics. I was fascinated by Euclidean geometry, then by calculus, but I didn't know what advanced research meant until I went to university in France.

In high school, I was subjected to the old French system with drills, problem solving, and tough, frequent exams. I was considered good at it—at least relative to my classmates—and got a scholarship to study in Paris. But, when I got there, I had to deal with a new reality. Some of my graduate contemporaries were simply amazing, enough to let you doubt your abilities and wonder whether you are good enough for such a career. Of course, it was much later that I realized that these graduate students were to be the world's best, such as Fields medalist Jean Bourgain, but also Michel Talagrand, Gilles Pisier, Bernard Maurey, and other future stars. Luckily, I stuck with it, and worked very hard to be able "to simply belong". I say luckily because now I realize that mathematics essentially saved my life (literally). It got me to Canada and earned me the wonderful life and family I now have.

AB: Who were the mathematicians or teachers who inspired you most in your career?

NG: For my last year of high school (equivalent to grade 12 or 13), I had to leave my mountain village to join a public school in Beirut. That was a first reality check. I thought that "I was good", until I met this amazing non-conventional teacher. I still remember his name (the only one I remember from high school): Moufid Said. The guy challenged us, provoked us, and essentially insulted us whenever we erred in our mathematical reasoning. He would state a problem on the board, and we had to race each other to solve it. For him, it was competition. His teaching methods defied any pedagogical theory, even then. First, we hated it; then it started growing on us. These were the late sixties. A long time ago! I was only fifteen then. That's when I started enjoying problem solving.

As with most families within my community, parents wanted their children to be physicians, lawyers, or engineers. I was preparing to get into medical school, when I randomly encountered my very same high school teacher, Said, on a Beirut street. He asked me about my plans, and I told him I was going into medicine. And here he was again in his overbearing, take no prisoner way: "If students like you are not going to do mathematics, I don't know who else would do it in this country." To the chagrin of my parents, I went to do a Bachelors of Mathematics. I sometimes wonder how many of our life-changing decisions are really ours, and how many are byproducts of random circumstances.

In France, I was inspired—and awestruck—by celebrities like Laurent Schwartz, and my supervisors Gustave Choquet and Antoine Brunel. They were giants in their respective fields then, members of the fabled French Academy of Science, and the whole works.

AB: You've worked in a variety of areas ranging from functional analysis to non-linear analysis and partial differential equations. Would you give us a summary of your current research focus?

NG: I seem to be as restless in my mathematics as in other aspects of life—whatever this means. My areas of interest evolved many times, influenced either by colleagues or by the pull of interesting problems. I started in functional analysis, which is somewhat a conceptual field, and then I switched to partial differential equations, which is a very technical field. Lately, I have been drawn to the theory of optimal mass transport, which essentially combines both features.

This theory started with a problem formulated by Gaspard Monge, who was a chief engineer in Napoleon's army: What is the optimal (least costly) way to transport a pile of rubble from one place to another. This mundane-looking problem has had a fascinating history spanning more than two centuries, including a major contribution by St. Petersburg mathematician, L. V. Kantorovich, who applied it in economics for which he won the Nobel Prize in 1975.

Another Russian, V. N. Sudakov, announced a proof in the sixties, which was eventually found lacking in the 2000s. The field is very active, with people considering various cost functions for different applications. For example, the case when the cost is proportional to the square of the distance travelled turned out to be very seminal in fluid dynamics, for functional inequalities, for differential geometry, and for

Figure 2. Gaspard Monge (1746–1818). (Wikimedia Commons, public domain.)

PDEs. My most recent interest is in the so-called Optimal Martingale Mass Transport and its applications to financial mathematics.

AB: You have been one of Canada's top mathematical leaders with involvement in PIMS, Mitacs, and MPrime NCE, and of course as Director of BIRS. What drives you forward in these ventures?

NG: In my first fifteen years in Canada, I was just a "normal" faculty member. By "normal", I mean minding my career, very possessive of my research time, avoiding committee work, travelling the globe to work with my co-authors all over. In 1994, I joined an NSERC grant selection committee, and for the first time, I started interacting with other mathematicians in different parts of the country, learning how the system works, and how mathematics was treated at NSERC compared to other disciplines. That's when I realized that there was a lot of work to be done in Canada and for Canada.

For one thing, we had many excellent Canadian mathematicians, yet limited resources and opportunities for them, mostly due to various states of ignorance by university administrators, industry leaders, and government agencies, as to what mathematics can do. There was also a big gap in opportunities and resources between Western Canada's mathematical science community scattered in a large and relatively isolated geographical area, and those in Southern Ontario (served by the Fields Institute), in Montréal (supported by the CRM), and with easy access to the mathematical centres in Boston, New York City, and Chicago. We wanted to link the scattered universities in our vast region to the rest of the (mathematical) world. That was the idea behind the creation of the Pacific Institute for the Mathematical Sciences (PIMS).

I was very green, but I had enthusiasm and was fortunate to have a wonderful group of colleagues to work with: Ed Perkins at UBC, Reinhard Illner at the University of Victoria, Claude Laflamme, Peter Lancaster, and Michael Lamoureux in Calgary, Nicole Tomczak-Jaegermann, Bryant Moodie, and Bob Moody at University of Alberta. Arvind Gupta, the late Jonathan Borwein at SFU [Simon Fraser University], and his brother Peter were a huge influence. As you wrote in your blog, Jon was a major asset: a great source of innovative ideas, energy, and political savvy.

PIMS was probably the very first distributed institute in the world. That was then a futuristic idea, only made possible by a recent advent of e-mail communications and the internet. We were trying to create meaningful links between mathematical scientists in a geographic area that was bigger than Western Europe (with BC, Alberta, and Washington state). We also connected mathematicians with the industrial and educational sectors. Now that looks like obvious things to do, with PIMS being one of the most successful mathematical science institutes in the world, encompassing fourteen universities and two countries. But these were novel ideas for that time, and funding was hard to come by. I have a whole shelf in my office of unsuccessful proposals to NSERC, BC, and Alberta governments between 1995 and 1999.

As soon as NSERC awarded PIMS a reasonably functional grant in 1999, I felt that with Fields and CRM, we may now be well positioned to apply for a Network of Centres of Excellence (NCE). The idea looked bizarre then, for many reasons. For one, our proposal pitching mathematical methodologies to address societal issues was to compete for government funding in "the big league" of high-profile, disease curing, deep universe exploring, big science projects. The network was to have a different mandate than the institutes, including getting the private sector to realize that mathematics was important for their bottom line. We were fortunate that Don Dawson directed the Fields Institute at the time, and Stephen Halperin chaired the Mathematics Department at the University of Toronto: two close friends who cared deeply about the future of Canadian mathematics. Together, we worked on the first proposal for Mitacs, and the rest is history.

AB: Like many mathematicians my age or younger, I grew up with BIRS and the institutes. But it's fascinating to hear how they all came to be.

NG: In the fall of 1999, I went on sabbatical to France to rest from all this, and to get my research back on track. I had been to Germany's Oberwolfach before, but it was while visiting CIRM [Centre International de Rencontres Mathématiques] in Marseille that I had the idea to develop a similar Centre in Canada. This was a bizarre idea then since a panel commissioned by the US National Science Foundation had asserted a year earlier that there is no need for such an institution in North America. I begged to differ.

Besides falling for the spectacular geographic location of Banff, and the place of high culture that the Banff Centre represented, I felt that Alberta was the ideal province to host a major mathematical Centre with a mandate that complements Mitacs and the other institutes. After all, Alberta's government, its mathematicians, university administrators, and community leaders had played a key role in establishing PIMS and Mitacs. And sure enough, thanks to a great Alberta science and innovation enabler, Bob Church, the provincial government was the first to pledge funding support. On the other hand, I felt the time was ripe to have a true partnership with the US, both on the operational and institutional level. So, I contacted my friend, David Eisenbud, the Director of MSRI in Berkeley, who quickly embraced and supported the joint initiative. Within six months, BIRS was funded! PIMS had taken essentially five long years, and here, lining up the support of three granting organizations serving three different governments (Alberta Innovation, NSERC, and NSF) took just six months!

Figure 3. The Banff International Research Station. (Photo licensed under Creative Commons Attribution 3.0 Unported (CC BY 3.0).)

And here, I must give lots of credit to Tom Brzustowski, then president of NSERC. Tom is a mechanical/aeronautical engineer, a former provost of Waterloo, who had a clear understanding of the role of mathematics in scientific discoveries and engineering innovations. He proceeded to break every taboo of the NSERC bureaucracy by asking his Vice-President Nigel Lloyd to accompany me to meet the NSF officials in Virginia, and carry NSERC's message: Canada is open to scientific collaboration between the two countries. I say taboo because NSERC's modus operandi is foreign to the notion of initiating, incubating, or even expressing early support for research initiatives. It is focused on reviewing and potentially funding formally submitted proposals via rigid bureaucratic processes. This feature is very relevant to current discussions in Naylor's review panel on how to fund "big science" in a Canadian context. How to initiate big multinational, multidisciplinary projects, and who speaks for Canadian science on the international stage?

Mexico joined the partnership a couple of years later, and more recently CONACyT provided major funding to build and run Casa Mathematica Oaxaca (CMO), which now hosts some of BIRS research activities. So BIRS was a major development on many levels. It is the first real and ongoing scientific partnership between Canada, the US, and Mexico. It is truly international, both operationally and institutionally, and everyone "pays their share", so to speak. The NSF covers US scientists, CONACyT supports the BIRS programs in Oaxaca, while European and other countries contribute substantially by covering the travel of their citizens to Canada and Mexico. The four granting agencies review BIRS on a regular basis by organizing joint site visits. We just had another successful one, and I am happy to announce that the funding is secure for another five-year cycle.

AB: You've been an outspoken critic of university administration, in particular with regards to events surrounding university presidents and boards.

NG: Sometimes I wonder if the biggest mistake of my life was to serve on the Board of Governors at UBC. I was there for six years (2008–14). Once you are there, you are exposed to how decisions are taken, how hundreds of millions of dollars are spent, how mistakes (honest or not) are made. Universities are becoming huge, profitable businesses here and in the US. And whenever you have large sources of cash, the vultures start circling: construction companies, developers, IT companies, energy outlets, athletic gurus, and the emerging merchants of sustainability and the green economy. It is incredible. As a mathematician, you can reason, compute, scale, and estimate.

So, you start being exposed to some unpleasant things. And once you get to know so much, you cannot ignore what is going on around you: the good, which is welcome, but also the bad and the ugly, which give you heartburn and sleepless nights. I sometimes wish that I kept the blissful ignorance of a regular faculty member—at least of that side of a university. It was a great learning but painful experience.

Figure 4. UBC, Point Grey, and the Strait of Georgia. (Photo from Shutterstock.com.)

AB: What do you think the role is of faculty in holding university administration accountable?

NG: I, like most faculty members, have my own vision of the university as a place of research and learning. That's why the university was created in the first place. The curse of UBC was that its president has to also be the mayor of "University Town". When you sit on the most expensive real estate in the world, this creates lots of opposing pressures: Are we a university or a Club Med for students on the Pacific? And when you approve capital projects, you have to make choices between maintaining old and decrepit academic buildings or building state-of-the-art aquatic centers, stadiums, tennis and squash courts, or commons.

I know you must be alluding to the short presidency of our colleague, Arvind Gupta. Well yes, I was involved in that search committee having been elected to represent the faculty. UBC has so many resources, and, if used right, we could lift the university and all of Canada to great heights. But we needed academic priorities to come first. Gupta's agenda was to refocus on the academic mission, so we had high hopes.

Things went bad for various reasons, and I knew more than most because, for one, I had mingled with the main characters on the Board, the Executive and the Deans, including those responsible for the president's resignation. I didn't think it was fair either for UBC or Gupta. But I became an outspoken critic, and I blogged about it, mostly so that such an institutional failure never happens again. The university governance needed to be reformed and taught to Board Chairs, Chancellors as well as Deans. How much can the Chair of the Board—who doesn't necessarily know about academic matters— interfere with the working of a university? Should the Board micromanage the president? Can the Deans make end runs and turn to Board members whenever they disagree with a president's decision? It's really about the future of UBC. If we don't fix these governance issues, then no president will be able to preside. The faculty revolted, and 800 of them eventually voted non-confidence in the Board. That was huge, even if they are trying to ignore or downplay it. The outcome of all of this is that we, the faculty, strengthened the presidency, which duly represents the academic side of university governance. Ironically, there is traditionally an "us versus them" mentality between the administration and the faculty. Here, we had a reverse situation, where the faculty supported the role of the president to make tough academic choices and define the priorities. Because of that, we now have Santa Ono, a president with a strong mandate. We expect him to use it, to reform, to refocus on the academic mission. But for that, he needs a major sweep of all the characters involved in last year's fiasco. Otherwise, he would be rewarding those who trampled on university governance and stalled UBC for many years. We are hoping he will do that sooner rather than later.

Figure 5. Santa Ono. (Photo licensed under Creative Commons Attribution-ShareAlike 4.0 International (CC BY-SA 4.0).)

AB: I think it's important for a president to have an academic background. They should be members of the academy.

NG: Absolutely. That's a given—at least for me. But unfortunately, I've also seen our academic colleagues sometimes going into administration and adopt that us versus them mentality. The best way is to work with faculty and work with them as partners. Faculty can be easily disarmed—if you will—once they know you are looking out for the best of the university. Santa Ono is so far doing very well in this direction. Arvind was a reformer. He may have proceeded too fast, and perhaps he didn't have a deep understanding of university politics, of historical entitlements, power brokers' expectations, and turf borderlines. But the painful episode exposed the risks of having non-academics or/and those unfamiliar with the academic mission

run the show at our universities. The Board is currently reforming
its governance structure, and I hope these issues will be front and
center in their deliberations.

AB: You are active on social media: Facebook, Twitter, and your
influential blog *Piece of Mind.*

NG: I came to blogging in a random way. I was sick and bored in bed,
so I started typing on my laptop, and I never stopped. But in the back
of my mind was also the fact that I owe it to the faculty who elected
me to the UBC Board to tell them what the big issues are facing the
university and my position on them. Then some NSERC issues arose,
and I became vocal about that too. Eventually, and perhaps because
I was saying aloud what many Canadian researchers were thinking,
the blog became very popular, attracting thousands of hits for each
post. I heard that political parties were distributing my posts in
their weekly briefings to their parliamentarians, especially on things
related to NSERC or government funding for research. I realized this
is useful! It surely was, especially when governance issues exploded
at UBC. However, it's a big responsibility and a huge investment.

I joined Twitter to advertise my blog. Facebook is more recent:
my daughter opened an account for me, and people I knew forty
or fifty years ago started to reconnect with me. Every medium is
interesting in its own right. President Ono is all over Twitter. And
I bet his social media blitz is putting pressure on Deans and other
presidents to join. Otherwise, they will look as secretive, isolated,
aloof, and non-transparent as they've ever been. I think this is going
to become the new way for accessing administrators, but also for
holding them publicly accountable. I think it is positive. The only
problem is how to manage time. You have to limit and control this.
If you solve this problem, let me know!

AB: What do you think is the importance of blogging and social
media for mathematicians in the 21st century?

NG: It's important to communicate to inform and involve the new
generation. Hopefully, they will take up the baton. My advocacy

stems from my own belief that mathematics is one of the deepest, most powerful and useful research disciplines. I am absolutely convinced that the depth and breadth and sophistication of mathematical research is second to none. This is my starting point. Unfortunately, it is not so for many—scientifically illiterate—administrators and bureaucrats. Try to explain the Langlands program. I want our community to be more confident of its business, and I want it to show its self-confidence. Without this view, mathematics will not thrive in Canada or the world.

Mathematicians should also learn about, and compare their resources to those of, other sciences. Young researchers need to know for example how, owing to inflationary pressures, BIRS had asked NSERC for an additional $75,000 above its previous grant. That was rejected because the math envelope is frozen. It has been for years. 2,200 scientists converge on BIRS each year, and an additional 800 go to Oaxaca (thirty percent of those are Canadian). The federal government provides the fourth lowest fraction of the BIRS budget ($2.6 million per year) after the NSF, Alberta, and CONACyT. They need to compare this to what transpired with the recent CFREF [Canada First Research Excellence Fund] announcements: more than $1.2 billion given to a dozen projects. There is something wrong in this picture, and a new generation of researchers has to get involved to get government and decision makers on their side.

Here is another one of our challenges as a community. As soon as a mathematical line of research becomes useful to another discipline or to innovators, it stops being called mathematics. The first generations of computer scientists were all mathematicians. They founded the first computer science departments in the late sixties. Now CS is a giant discipline that dwarfs ours. The whole new economy is based on "mathematical instruments", but uses different names to brand those: analytics, cryptography, artificial intelligence, and data science. All of these are offshoots of mathematics. The same goes for the so-called pure mathematics. Think of how topology and its byproduct, quantum topology, enabled the recent Physics Nobel winners. Watching the chair of the Nobel committee explaining topological invariants with pretzels and bagels must be a familiar sight to many of our undergraduate students. They know, but most others don't. Social media should allow them to spread that gospel.

AB: I'd like to close with looking forward. As Director of BIRS, you have a bird's-eye view of many breaking trends in mathematics. What would you say are some of the major directions for mathematics in the future?

NG: It is mind-boggling to see so many emerging areas of mathematical research developing at such an exponential rate. As of last week, BIRS received a record 210 proposals for its 2018 program. Now we have thirty outstanding mathematical scientists on the BIRS review panel, yet you would be surprised how many blind spots we have: new mathematics and new applications that none of us had heard of before. All of science is evolving at a rapid pace, but mathematics is at the forefront of this race for knowledge. To answer your question, I have first to mention the remaining classical big mathematical questions popularized by the $1 million prizes of the Clay Institute. It is remarkable to me that the Fermat and Poincaré conjectures have been solved in my lifetime, but not the regularity of solutions to the Navier-Stokes equation or the $\mathbf{P} = \mathbf{NP}$ question. Is it because the answer to the latter problems will turn out to be negative? I believe that the Riemann hypothesis is the Holy Grail and is in a class of its own.

But there are many—less classical and even more challenging— mathematical issues that will be occupying the next generations of mathematicians. I say "issues" because they are not distilled yet to the level of becoming specific "problems". The most frequently mentioned query is whether mathematics can be as successful in biology as it has been in physics; for example, what are the Fundamental Mathematical Laws of Biology? In the same spirit, can we develop a mathematical theory to build a functional model of the brain that is mathematically consistent and predictive rather than merely biologically inspired? Not unrelated is Mumford's call for new mathematics for the 21st century that captures and harnesses "stochasticity in Nature". Finally, the advent of powerful computers and the need for more powerful computational power has again placed mathematical techniques in a central position in science and technology. This is particularly true of the computational and algorithmic aspects of mathematics, but it pertains as well to techniques of modelling, analysis, and simulation. Mathematics is an amazing human endeavor that is at the very core of our quest for understanding, discovery, and progress. Most people are still not aware of it, and that's why we have to keep telling our story.

Chapter 8

Interview with Lisa Jeffrey

I met Lisa Jeffrey in Ottawa, where we worked together on the NSERC Evaluation Group for Mathematics & Statistics. Lisa comes off as modest and reticent, which reminds me of the quote by Stephen Hawking: "Quiet people have the loudest minds." We walked back to the hotel from dinner one evening and discussed her field of symplectic geometry, which was largely a mystery to me. She described the symplectic camel, and I knew then I had to learn more from her.

Lisa is a professor at the University of Toronto whose research focuses on symplectic geometry and mathematical physics. She is highly acclaimed, winning the Krieger-Nelson Prize and the Coxeter-James Prize from the Canadian Mathematical Society. She gave a prestigious Noether Lecture this year. Lisa is also a Fellow of the American Mathematical Society and a Fellow of the Royal Society of Canada.

This interview was conducted in December 2017.

$$\infty$$

AB: Where were you born and what did your parents do?

Figure 1. Lisa Jeffrey. (Photo courtesy of the University of Toronto, Scarborough.)

LJ: I was born in Fort Collins, Colorado. My (Scottish) father was doing his Ph.D. in forest hydrology at Colorado State University. My (Canadian) mother also worked in forestry and forest pathology—she now is very much involved in the environmental movement.

My parents met in Calgary working for the Canadian government. She became research officer, which at that time wasn't something that very many women did. We moved to Canada six months after I was born.

AB: What is your first mathematical memory? That is, a time in your youth where you had a vivid memory of something related to math.

LJ: In grade 8 or 9, I did a project on the Goldbach conjecture (which states that every even integer greater than two can be expressed as the sum of two primes). I just stated that it was. I obviously didn't prove it!

Another mathematical memory came from the time we lived in northern Norway for a year and a half when I was nine and ten. My stepfather got a job there. I was in the Norwegian public school, and my mother arranged that I would take math classes at a grade level two years higher than my own. The teacher (Mr. Saeboe) was very helpful and encouraging. I was in grade 6 math, and there was an exam at the end of the year and, apparently, I got the highest score of anyone in the county. This is not a large-scale achievement (Norway is a small country—population four million—subdivided into twenty counties), but maybe this went to my head.

AB: How did you decide to choose mathematics as a major in university?

LJ: I was a physics major, and I switched to mathematics in graduate school. After having done a physics undergraduate degree, I was awarded a Marshall Scholarship for study in the UK. I did parts II and III of the Mathematical Tripos at Cambridge. That brought me up to the level where I could think about continuing with mathematics. If you are a physics major, then you take quite a lot of math anyway, but I wouldn't have been equipped to start a math Ph.D. without my two years at Cambridge.

In my undergraduate years, I took real analysis, complex analysis, algebra, differential equations, and a course on Riemann surfaces; I took two courses on mathematical physics that were basically analysis (we followed a text by Ivar Stakgold which dealt with topics like the Fredholm alternative).

There were posters of historical mathematicians in the undergraduate physics and math library when I was an undergraduate. There were two female figures: Sofia Kovalevskaya and Emmy Noether. Both of them were role models for me.

Figure 2. Sofia Kovalevskaya (1850–1891) and Emmy Noether (1882–1935). (Wikimedia Commons, public domain.)

AB: How did you come to be Michael Atiyah's doctoral student? What was his style of supervision?

LJ: We met weekly, and he always had an hour allocated for each student. There were always many ideas from him, and he provided a lot of starting points. By the way, Ruth Lawrence was two years ahead of me (also working with Atiyah), although she was six years younger than I was.

In my first year, Ed Witten had just written his paper on quantum field theory and the Jones polynomial. There was a seminar in Oxford that fall about this material. Ruth Lawrence edited the notes, and quite a lot of material in Atiyah's book "The Geometry and Physics of Knots" was based on that seminar. That was the point of departure for my thesis: I was working on Chern-Simons gauge theory.

AB: What is symplectic geometry/topology?

Figure 3. Sir Michael Atiyah. (Photo licensed under Creative Commons Attribution-ShareAlike 2.0 Germany (CC BY-SA 2.0) DE.)

LJ: One of the standard examples people use to describe the field is the symplectic camel. Anyone in the Christian tradition will have heard the quote from the Bible: "It is easier for a camel to go through the eye of a needle than for a rich man to get into heaven." Now imagine you have a ball on one side of a plane and there is a small, circular hole in the plane. How would you get the ball through the hole and onto the other side?

If it were a question of volume, then you would squeeze out the ball and make it long and narrow. Then you would thread it through the hole. But to preserve the symplectic structure, that's not good enough. The radius of the ball would have to be smaller than the hole's radius. This example is discussed in the book "Introduction to Symplectic Topology" by McDuff and Salamon. Symplectic structure is the natural mathematical home for classical mechanics. It is the natural home for Newton's laws of motion, which can be rephrased as Hamilton's equations.

Noether's theorem (for any symmetry there is a conserved quantity—for example, symmetry under rotation corresponds to conservation of angular momentum) goes back to Emmy Noether and is a fundamental principle in symplectic geometry and Hamiltonian mechanics.

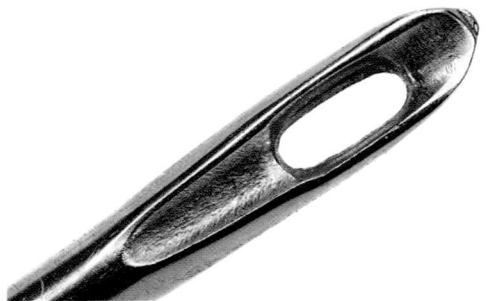

Figure 4. In symplectic geometry, you cannot squeeze a ball through the eye of a needle unless the radius of the ball is small enough. (Photo from Shutterstock.com.)

AB: What are you working on now?

LJ: The fundamental group of a space is the set of loops in the space, where you can deform the loops but not cut them. For example, the plane has a trivial fundamental group as you can always shrink any loop to a point. If you puncture the plane, then the fundamental group is no longer trivial, as a loop around the puncture cannot shrink to a point. Basically, the fundamental group is classified by the winding number, which counts the number of times the loop goes around the hole. So, the fundamental group of the punctured plane is the set of integers, counting the number of times and which direction you wind around the hole.

The space I've worked on is representations of the fundamental group into some other group such as the circle group. In the case of the punctured plane, this would be one copy of the circle group as you just have to say where the generator of the group goes (it goes to some point on the unit circle). I've worked on other more complicated examples that come up often. There was a groundbreaking paper of Atiyah and Bott in 1982 where they studied the space of representations of the fundamental group of a 2-manifold. A 2-manifold would be a torus or a 2-dimensional sphere, or anything you get by gluing these structures together (classified by the number of the holes, which is called the genus).

AB: Does your research interact much directly with physics?

LJ: I published a paper about five years ago that was from my Ph.D. thesis that did have to do with physics. As a result of that, I spoke at a physics conference Theory Canada 9 at Wilfrid Laurier University in Waterloo. Many of the talks there were straight physics. In hindsight, it would have been better if I had rephrased my talk in physics-language.

AB: I've been thinking a lot lately about diversity in mathematics, and why women constitute only about twenty percent of mathematics departments in Canada. What are your thoughts on this? How can we change this culture?

LJ: In our department, fifteen percent of our faculty are women, and the same percentage holds among our graduate students. I wish the numbers were higher. I don't understand the graduate percentage, as often our admissions committees are composed of women. It's not a matter of any discrimination, but we just have fewer women applicants. I don't think our field has as much gender disparity as in engineering or perhaps physics. How to change the culture? When I was in high school, there was a lot of attention paid to the question of why girls were dropping out of math class and what to do about it. Now there is much discussion of how the school system is failing boys. There are books on the "war against boys". So, somehow, the focus has changed.

Many people think that women are over-represented in universities, but this is not true in STEM fields. Medical and law schools

will typically have more women than men. People need to be re-
minded that there are still issues getting women to study STEM.
The problems in the education system encountered by women have
not disappeared.

Figure 5. Matilde Marcolli. (Photo courtesy of the Archives
of the Mathematisches Forschungsinstitut Oberwolfach.)

By the way, the University of Toronto and the Perimeter Institute
just hired Matilde Marcolli from Caltech. This is fabulous news.
Matilde is going to be a role model, splitting her time between both
Toronto and the Perimeter Institute.

AB: What is your advice to young people (especially young women)
who are considering studying mathematics at university/grad school?

LJ: The important thing is to understand that math leads in many
directions, not just academic ones. People with mathematical training

will be highly employable. One of my best students finished his Ph.D., got a post-doc, and within eighteen months he had a programming (non-academic) job at a physics research institute. He had no trouble at all getting a good job.

AB: I always close looking forward. What would you say are some of the major directions for mathematics in the future?

LJ: There are many questions related to the work of Nigel Hitchin on Higgs bundles (the same Higgs associated with the Higgs boson). That work has major ramifications in many different directions, including to representations of the fundamental group that I discussed earlier. That work is an outgrowth of the 1982 paper of Atiyah and Bott, but it takes things in a different direction.

Chapter 9

Interview with Izabella Laba

Izabella Laba is a leading mathematician working at the University of British Columbia. Her research spans several fields such as harmonic analysis, geometric measure theory, mathematical physics, and the relatively new (but fast-breaking) area of additive combinatorics.

She is highly awarded, having won the prestigious Coxeter-James and Krieger-Nelson Prizes of the Canadian Mathematical Society. In 2012, she became a Fellow of the American Mathematical Society. Izabella was an invited speaker at the International Congress of Mathematicians in Seoul in 2014.

Izabella is one of the small, but growing, set of mathematicians active in research and also actively engaging social media. Her blog *The Accidental Mathematician* has great pieces on her experience with mathematics, gender, and the academy. Her fearless candor has greatly influenced me and many others. She is an avid photographer, especially of nature, and she maintains an archive on Google+ of her photos.

Figure 1. Izabella Laba. (Photo courtesy of Izabella Laba.)

This was my first interview with a mathematician outside of my field of networks (although I also work in combinatorics, but in a different direction than hers). Izabella was, as I expected, both insightful and generous with her responses. I've never heard mathematics compared to architecture before, but the analogy is an excellent one.

This interview was conducted in September 2016.

∞

AB: When did you first realize you wanted to be a mathematician?

IL: That was more like a process than a realization. It was never a clear-cut choice. I've always been interested in mathematics, but I also always had other interests. Part of it was that mathematics was how I got out of Poland; I was admitted to graduate school at the University of Toronto. That happened at a time when I was considering other options because my mathematical work was not going anywhere. It was a sequence of such choices that led to the career I have now.

AB: Who most influenced your early career?

IL: I'm not sure that anyone did. My mother encouraged me in mathematics, but not always in ways I appreciated. On the other hand, I also had teachers who discouraged me from mathematics, and that made me want to study it more.

I was a good student, but I was female, and that was a problem. There was this one teacher in high school who said in class, and also told my parents in private, that girls did not have research careers. They might be good in school, but then they would just get married, and that would be the end of their intellectual pursuits.

AB: You've proven them wrong!

IL: I guess so.

AB: You completed your Masters in Poland, and then a Ph.D. at the University of Toronto. What was it like splitting your graduate work across two countries?

IL: I had to learn English. I had to learn some mathematics again in a new language. I had to get used to living elsewhere.

One of the hardest parts was just applying to graduate school. Poland was still a communist country then, and communication with the rest of the world was very difficult. For example, I would get a scholarship notification letter in the mail three weeks after it was sent, and then I would only have one week left to respond. With the Polish postal service barely functional at the time, making the deadline required serious effort. There were issues like that all the time, making the logistics extremely complicated. Studying mathematics had been easy by comparison.

It took me some time to get to a point where I could do actual research. I had setbacks and projects that didn't work out. I did not solve the first question that was supposed to be my thesis topic, on asymptotic completeness for long-range multiparticle systems with an improved range of exponents.

AB: That was a question in mathematical physics?

IL: Yes. It was very difficult and is still unsolved.

AB: You work in many areas such as mathematical physics, analysis, geometry, number theory, and combinatorics. How did you come to work in these areas?

IL: Here I should mention someone who did influence my mathematics: Tom Wolff, who was at Caltech at the time when I worked at UCLA. I met Tom there, started attending his seminar and got interested in his mathematics. This was also a time when I began looking for ways to break away from mathematical physics because I didn't want to be in that field for the rest of my career.

Tom's mathematics appealed to me as it pursued questions instead of fields. Instead of working in this or that field, you went after a question, and you went wherever that question took you. If it meant that you had to learn some combinatorics or number theory, then you just had to do it. I found that exciting and different from what I had been doing up to that point. It really appealed to me.

AB: Which of your research accomplishments are you most proud of?

IL: Probably at any given time, it would be my most recent paper. Right now, I just finished an article with Hong Wang on multiscale lambda-p sets. These are fractals that are very highly random, in

Figure 2. Tom Wolff (1954–2000). (Photo courtesy of the Oberwolfach Photo Collection, licensed under GNU Free Documentation License, Version 1.2.)

ways that can be quantified using harmonic analytic estimates. We use a lot of harmonic analysis, geometric measure theory, and some probability. It was fun to work on it.

I've had several projects along similar lines, where my collaborators and I take results from harmonic analysis and try to develop analogues for fractal sets that might be of interest in geometric measure theory or dynamical systems. This links several areas of mathematics. On the one hand, there's inspiration from many different directions. On the other hand, it's not always straightforward to move ideas from one field to another, so then you have to develop something new.

Another project that I had a few years ago was on Favard length of sets. There, you start with a problem in geometric measure theory, but if you're interested in quantitative results, you need to do some harmonic analysis. And then it turns out that the harmonic analysis part also requires some number theory. We even had to dig up some old combinatorial results about sums of roots of unity.

AB: Academic publishing is changing rapidly. You are active in the new arXiv overlay journal *Discrete Analysis*. Please tell us more

about your involvement with the journal, and your perspective on it
in contrast to traditional academic publishing.

IL: I'm on the editorial board of the journal. When a submission is
received, there is usually a quick conversation to see if we want to
proceed with the detailed refereeing. If so, then the editors assign
referees, and there is another discussion after reports are received. If
a referee suggests revisions, we supervise the process.

Unlike conventional journals, we don't do copy editing or for-
matting of the paper. There are no subscriptions or printed physical
copies of the journal. Our web page aggregates links to the papers
and provides short editorial descriptions, but the papers still live on
the arXiv.

AB: This is the first mathematics journal I am aware of that is an
arXiv overlay.

IL: There were other such journals previously, but not in mathematics.

AB: Do you think this is a trend for the subject?

IL: It might be. I can easily see other mathematics journals like
that being created, and I'd be happy to have more of them. I don't
see overlay journals replacing the traditional publishing system altogether, but they could become a significant part of academic publishing.

AB: What inspires your mathematical ideas?

IL: That's hard to describe. Much of it comes from reading other
people's mathematics. You find an idea here, and you think about
applying it elsewhere or combining it with a different idea you've
found somewhere else.

I need intellectual inspiration beyond mathematics, even if it's
indirect. I would not go so far as to say that this painting or that
newspaper article inspired this specific mathematical idea. But I do

know that if I am deprived of stimulation like art or reading, then my mathematics suffers.

AB: Mathematicians say they are drawn to the beauty of the subject. What role do you think aesthetics has in mathematics?

IL: There is a role for it, but it's not completely clear what it should be. It might be a little bit similar to architecture, in that beauty is tied to functionality rather than pursued independently. The building has to stand upright, it should support infrastructures like plumbing and the electrical system, and at the same time, it should also be pleasing to the eye. And there are also times where the architect's vision for what a building should look like inspires the development of new technical solutions to make that possible, so there's a give and take in both directions.

Figure 3. "The building has to stand upright... and at the same time, it should also be pleasing to the eye." Izabella Laba. (Photo from Shutterstock.com.)

Another example would be the design of electronic devices. Apple gadgets can be beautiful and pleasant to hold, but they also have to be functional, and their beauty is not separate from that. The form and the function influence each other.

Mathematics is a little bit like that. I don't set out to produce beautiful mathematics. I want it to be correct and to answer questions that people might be interested in. Making it appealing and beautiful can be directly at odds with the requirement to keep it correct and complete. I try to reconcile these competing demands where possible, and that often improves my work. But where they can't be reconciled, I have to stay with the correct and ugly.

AB: What advice would you give to young people, especially young women, who want to study mathematics?

IL: I'm not sure that there's any blanket advice that I could give to everyone. People are not all the same and what works for one person might not work for someone else. You see it very clearly when you work with graduate students, especially if you have more than one at a time. They all have different personalities, attitudes, approaches to mathematics.

Perhaps the one piece of advice that I could offer is to never discourage anyone from studying math just because they are different from you. They might not look like you, talk like you, or do the same things as you in their free time, but none of it means that they can't do mathematics. By the same token, if you are different from other mathematicians you know, that doesn't mean you shouldn't be there.

AB: You are active in social media through your blog and tweets. What do you think is the role of social media in the mathematics community?

IL: I don't think of my use of social media as connecting with the mathematics community. I don't know if there even is a mathematics community! There are mathematicians—whether it's a community, that's a different question.

AB: Do you think your social media interactions can help people understand mathematics better?

IL: I interact with many people on social media, not just with mathematicians and not only about mathematics. Now, these people may be aware that I am a mathematician. If they imagined mathematicians as people who just sit in their respective ivory towers, not interested in talking with anyone else, then it's possible that my presence on the internet might contradict those ideas. But generally, I'm not out there trying to educate or proselytize. I choose the groups that I interact with based on things like shared interests or willingness to engage in a discussion. Mathematics is not a very large part of that.

AB: You maintain a Google+ page containing your photography. Can you talk about your photos, your subjects, and what interests you?

IL: I love taking photos of nature. I'm very lucky to live in the Pacific Northwest. I fell in love with this area when I first visited a long time ago. That's where I wanted to live, and that's where I live now. My photography is my way of processing how I see nature—how I think about it. Mostly it's something I do for myself, although I do post photos on the internet and I'm glad when people like them.

AB: Let's finish by looking forward. What do you think are the major directions in mathematics?

IL: If you are asking about specific developments or areas that might become hot, I don't know. If you are asking about general trends in how mathematics is going to develop, then I think we might see more interaction between different areas of mathematics. We might see new areas of mathematics forming, the way additive combinatorics formed over the last decade or two.

There is much more connectivity between people now, with more opportunities to talk to each other. There is more collaboration. I think the result of that is that people will be less likely to just stay in one area of mathematics, and more likely to branch out or move back and forth between several fields. This is already happening in science and the arts. I think mathematics develops more slowly (for

Figure 4. Photo by Izabella Laba. (Photo courtesy of Izabella Laba.)

example, it takes longer to write or read our papers), but the general direction is the same.

Chapter 10

Interview with Barry Mazur

Barry Mazur is the Gerhard Gade University Professor in the Department of Mathematics at Harvard University. Barry is one of the world's leading experts on number theory, and his work focuses on Diophantine geometry and elliptic curves. His research in topology is also legendary, and he settled the generalized Schoenflies problem as a doctoral student.

He is highly awarded, having received the Leroy P. Steele Prize for Seminal Contribution to Research, the Cole Prize in Number Theory, the Chauvenet Prize, and the Oswald Veblen Prize in Geometry. In 2011, he received the National Medal of Science from President Obama. Barry is also a Fellow of the American Mathematical Society and the National Academy of Sciences.

I met Barry in 2004 while visiting Fan Chung Graham in San Diego. He was attending a conference in honor of Persi Diaconis and Fan introduced us. I was initially intimidated meeting such a mathematical giant, but I recall him greeting me with a warm smile and politely shaking my hand. In the interview, he spoke thoughtfully and deeply, and he has an infectious laugh that immediately put me at ease.

This interview was conducted in May 2017.

Figure 1. Barry Mazur. (Photo author: Gert-Martin Grevel. Photo source: Archives of the Mathematisches Forschungsinstitut Oberwolfach.)

∞

AB: Did you show an interest in mathematics as a child or did that come later?

BM: I remember being puzzled when I was a very small child by patterns that have what might be called a "mathematical feel".

Here is an example, which I have written about. I was fascinated by this simple question: if you count the fingers on one hand, you have five, and if you count the crevices in that hand, then you have four. Whether that is mathematics, I don't know—it's a kind of elementary thinking about patterns that I'm sure everyone does. We look at a flower or a teacup, and we're struck by its symmetry or its engaging lack of symmetry—by any structure that could be thought of as geometric.

Figure 2. "We look at a flower, and we're struck by its symmetry." Barry Mazur. (Photo from Shutterstock.com.)

AB: Was there a family member or teacher, before your university education, who supported you to study mathematics?

BM: My father would always test me with little puzzles. I don't know how old I was, but I was appropriately young for this type of puzzle: what number, when you double it and add one, do you get eleven? My approach at that point was brutal trial-and-error. Frustrated, perhaps, by my experimental approach, at one point my father said, "I will show you a secret." He wrote at the top of a blank sheet of paper "Let x be the number when you double it and add one you get eleven." Then he carefully wrote out the rudiments of doubling x and adding 1, and then suitably unveiling the x to find that it is equal to 5. He was very fastidious about his instructions. I was both beguiled and happy with that. I cherished its secrecy as much as its effectiveness.

He would quiz me from time-to-time, and I would find answers for him, armed with our family secret. I was astounded, some years later, to find this very family secret is revealed on the blackboard to the entire math class by a teacher. Of course, it is rather its un-secrecy, the availability of mathematics to everyone, that we should press for!

AB: You completed your doctorate at Princeton in the 1950s. Would you tell us about the mathematics department at Princeton back then? How did you end up working with your supervisors there?

BM: Princeton was wonderful. At that time, it was vibrant with algebraic and differential topology. John Milnor and Norman Steenrod were there, as well as a number of other great innovators of various aspects of topology.

But more important to me than Princeton was MIT where I did my undergraduate studies. I arrived at MIT with a passion for electronics. What enthralled me about the subject is what one might call "the philosophical aspects of electronics having to do with action at a distance" and the issue of electromagnetism. I was deeply impressed by the amateur radio enthusiasts who I had met in my high school years. I thought that "radio waves", whatever they were, constituted a great mystery that I had to get to the bottom of: how is it that energy leaving the antenna of a transmitter somehow manages to find the antenna of a radio receiver? What is it doing between the moment it leaves and the moment it arrives? This was my motivation for going to MIT and trying to learn the mathematics I needed to understand this.

The first day I got to MIT, I discovered the library and the immense resource of books and journals, all about the very mysteries I wanted to understand. What I quickly realized, though, was that it was precisely the mathematics behind those mysteries that I was interested in. I immediately changed my emphasis from electronics to physics, and then to mathematics.

I was at MIT for two years and managed to get into the graduate school at Princeton in my third year. In graduate school there were, as I mentioned, wonderful teachers but not too many classes at the levels I was capable of absorbing (or, in fact, at any level). At that time, there were fewer classes than there are here at Harvard. I, like many other graduate students, was therefore left to my own resources. I remember going to an undergraduate course in Galois theory given by Emil Artin, and a graduate course in C*-algebras by [Irving] Kaplansky (who was visiting Princeton at the time). We graduate students would go to a course or two, but, otherwise, we would set up our own seminars to learn things. We cobbled together a hit-or-miss type of curriculum that probably wouldn't have been approved by the older generation. We studied some physics, books by Arthur Eddington, lots of point-set topology, and random topics. All of it was thrown together and very exciting.

I was there at Princeton for a year as a graduate student. One year is not that long a time, but still, the next year I wanted a break, so I went to Paris with a girlfriend from high school. There in Paris, I had the urge to try to prove the Poincaré conjecture. Paris was extremely lively back then mathematically (as it always is), and I went to a number of courses there, studying more algebra than I had previously done.

Of course, I didn't manage to prove the Poincaré conjecture, but I did prove something that I called Lemma 1, a lemma that I didn't think of as a significant step towards the Poincaré conjecture. I came back to Princeton at the end of the first year, forgetting about my Lemma 1. But one day, in the common room of Princeton I heard Ralph Fox, the great knot theorist, talking about various interesting open problems in topology. He mentioned something he called the Schoenflies problem, and I immediately recognized it as my Lemma 1. I brashly told him, "I can do that!" His response, as would only be natural, was quite dismissive: he said that if I could do that he would make me a professor at the Institute.

I wrote up my proof and showed it to Christos Papakyriakopoulos. He was a long-term visitor at Princeton and a great topologist. As I started to explain my proof on the blackboard of his (tiny) office, he said: "Prove it in dimension 4." But Lemma 1 works in every dimension, so I told him I could—without changing the words of my proof—prove this Schoenflies problem in any dimension n. He insisted, though, that I do it in dimension 4. In my explanation of my proof to him, every so often I would forget and say "n" rather than "4", and he would correct me: "we are in dimension 4" he'd say. Later, this turned into something of a running joke between us. He was gracious and supportive, but the number 4 plays no role in my proof.

Another great topologist, R. H. Bing, was at Princeton at the time, visiting the Institute for Advanced Study. I showed my proof to him and he, too, was incredibly gracious. He said I should give a lecture about this at the Institute, and explained to me among other things, how to give a lecture, and how to draw the diagrams I had to draw. Bing's mathematical work is magical, and it delves so deeply into the essence of three-dimensional topology. A striking example of this magic is his amazing theorem that—describing it perhaps a bit too succinctly—the double of the wild component of the Alexander horned sphere (that is, two copies of this component sewed together on their boundary) is homeomorphic to the three-dimensional sphere.

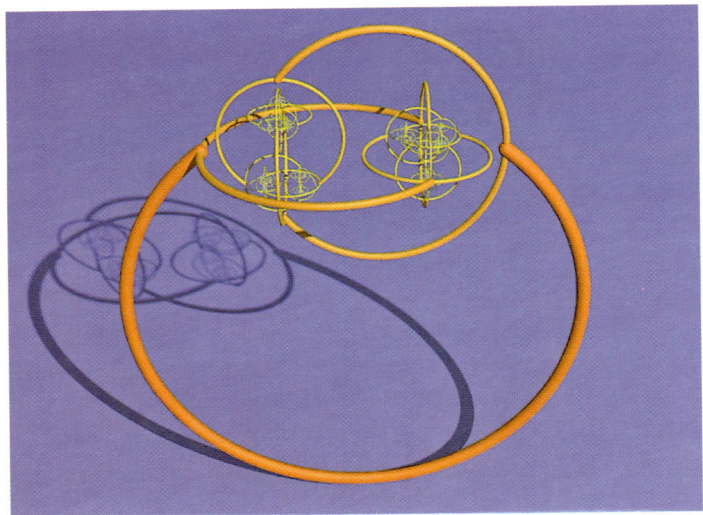

Figure 3. The Alexander horned sphere. (Wikimedia Commons, public domain.)

I brought the written version of my proof of Lemma 1 to Ralph Fox, and I said I wanted this to be my thesis. He didn't say much, but I did graduate and received an invitation from the Institute for Advanced Studies to study for a year as a post-doc (I assume that it was thanks to him). J. Robert Oppenheimer was the director of the Institute at the time and very generous with his time. He once drove me to the house of J. W. Alexander (the creator of the Alexander horned sphere) who was retired from the Institute but lived nearby. I vaguely remember that we did have a conversation, but I was quite tongue-tied.

AB: While your early work focused on topology, later you moved to work in number theory, especially Diophantine geometry and elliptic curves. Can you explain for the layperson how geometric or topological thinking influences the study of numbers?

BM: It's always shocking when you see two intuitions that seem to have nothing to do with each other somehow combine and reinforce one another—then synthesize to form some new powerful viewpoint. The names of some fields of mathematics already suggest this: a name with a noun and adjective combined like algebraic geometry gives you a hint there is some synthesis built from the combination of algebra and geometry. That happens more than one would imagine when you study mathematics.

One of the great links between topology and number theory is between knots (which are closed strings in three-dimensional space) and prime numbers. It would be difficult to explain what the analogy is without being technical; so I won't. But for me, that was a helpful springboard to move from topology to number theory—a way of passing from a topological intuition and merging it with an arithmetic intuition about primes. That is one link, and there are many others.

I was also interested in dynamical systems. There are many types of dynamical systems, but the type I was thinking about are "discrete dynamical systems". Such a dynamical system is based on a transformation T of a geometric space to itself. What happens when you perform that transformation again and again: for example, take a point

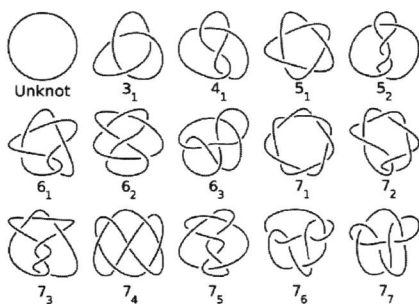

Figure 4. Examples of prime knots with up to seven cross-ings. (Wikimedia Commons, public domain.

x and see where it is sent to after iterations of that transformation—that is, $T(x)$, $T(T(x))$, $T(T(T(x)))$, ... You get an orbit. Such orbits are often fascinating; they often have some interesting topology, there may be attracting or repelling orbits, they may be densely distributed. Dynamical systems are very beautiful in their own right, but they also have incredible utility and application in disciplines such as physics and other subjects. It occurred to Michael Artin and me to apply algebraic geometry to a question in dynamical systems. For this, I had to learn some "real algebraic geometry" (the theory of Gnash manifolds), and that got me hooked, and I became more interested in algebraic geometry and, eventually, number theory.

AB: What research topics are you working on most recently? You can be more technical here if you like.

BM: I am interested in rational points, which are solutions of polynomial equations—the coordinates of these points being rational numbers (or fractions). Take an equation such as $y^2 = x^3 + 1$ which you can visualize as a curve in the plane. The problem is to find the full list of points in the plane that have rational coordinates and happen to lie on that curve. This type of problem has an extraordinary beauty but is also often important for applications or other mathematical pursuits.

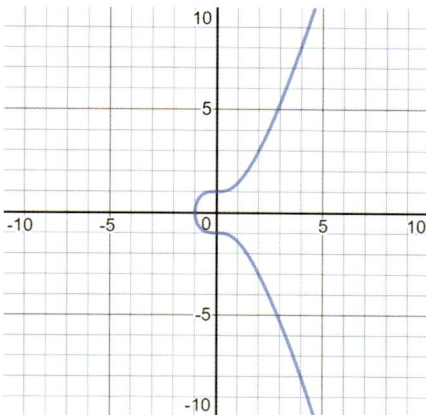

Figure 5. The elliptic curve $y^2 = x^3 + 1$.

I mentioned $y^2 = x^3 + 1$ as my example because, among polynomial equations in two variables, linear and quadratic polynomials have one type of behavior, high-degree polynomials quite another. There is a very curious interface—between these two different "types of behavior"—that occurs with polynomials of degree 3: you are in the land of elliptic curves. Elliptic curves are seemingly confined and specific, but they are ubiquitous in mathematics and show up in as many theories you can imagine such as complex analysis, number theory, lattices, and even in the study of the heat equation. In finite fields, elliptic curves become the foundation of coding theory and other aspects of the practical world. I'm told that every time you use your bank card, this involves elliptic curves in the manner in which information is encoded. The basic issues in pure arithmetic and number theory depend on understanding elliptic curves. Elliptic curves lead to deep pure mathematics and have important applications in coding and cryptography.

AB: Your latest book *Prime Numbers and the Riemann Hypothesis* presents these topics to a math undergraduate.

BM: Actually, an engaged high school student can read the book. It's not a passive read as we are inviting people to do our computations or do them in a different style. I also thought that engineers could well be interested in it.

AB: I've tried to explain the Riemann hypothesis to non-mathematicians with mixed success. How would you explain the Riemann hypothesis to someone who has a limited math background?

BM: There are steps in my expository game for this. The first is to make sure people understand the importance of primes. If you don't realize the primes are interesting, then the Riemann hypothesis is not interesting. The second thing to realize is that prime numbers have on the one hand such erratic behavior: 2, 3, 5, 7, 11, 13, 17, 19, ... They don't seem to have a clear mnemonic that will allow you to remember the first 100 primes, say, like there would be for the first 100 numbers divisible by 10. The primes seem to come randomly, and they keep coming (that is, there are infinitely many primes).

But then—and it is quite impressive—when you look at them from afar they seem to have such clear structure. Draw a graph (on one standard-size page) that charts the number of primes less than x, and where the range is for values of x between 1 and, say, 40, and it looks like the most randomly constructed staircase. And now draw one where the range is for values of x between 1 and, say, 10,000 primes and you get a strikingly smooth graph! Looking at primes from afar, their erratic turbulence disappears. That is the perplexity: from afar the graph of primes is so smooth, while from near it looks disorganized and complicated. The Riemann hypothesis continues the grand project begun by Gauss of explaining exactly how smooth that progression of primes is.

Another point is that the Riemann hypothesis is a type of conjecture that suffuses. It's not just nice to know the answer—but once we do know the answer, we also know much more; so many other pieces of mathematics depend on knowing the validity of the Riemann hypothesis.

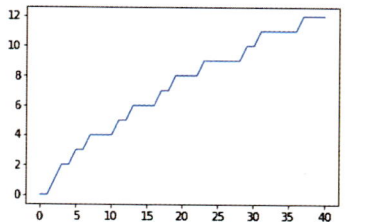 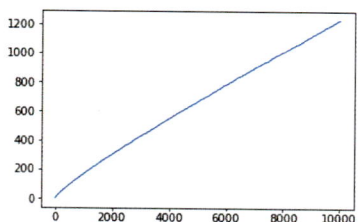

Figure 6. The number of primes up to 40 and up to 10,000. (Graphs courtesy of Narges Alipourjeddi.)

AB: You've had many students over the years such as Jordan Ellenberg. What is your advice for young people (say graduate students) studying mathematics?

BM: The first thing I hope people learn (this applies to both undergraduates and graduates) is to respect their native curiosity. When you are curious about something, then you can ask questions that matter to you, and that's extremely important. You should follow those questions. There is an art of asking questions in mathematics that you should cultivate. It will help you enormously.

Practice the art of understanding and respecting your own questions.

AB: I'd like to close with looking forward. What would you say are some of the major directions for mathematics in the future?

BM: Mathematics is too broad to predict. Overnight, there may be a new road. For example, I went to the Arizona Winter School, which chooses a subject (often in number theory) and has senior people giving lectures. There are usually many students, who do projects on the subject that is being taught. It's a week-long thing. Very often the subject is one that has just opened up.

This March they covered the topic of perfectoids. I won't tell you what it is as it is rather technical. But I knew I had better learn perfectoids, so I signed up. I didn't give lectures; I just went to learn.

Figure 7. Peter Scholze. (Photo author: George M. Bergman. Photo source: Archives of the Mathematisches Forschungsinstitut Oberwolfach.)

This is largely around the work of Peter Scholze who is extraordinary in many ways, and has developed a marvelous school of collaborators developing perfectoids. They have produced an enormous number of results in arithmetic algebraic geometry, arithmetic, and representation theory.

The first thing that struck me was that there were between 300 to 400 graduate students learning perfectoids. It's just as it should be, as this new development opens up an area that is extremely important. The lectures and exercise sessions were really good. It was one of the most exciting winter schools I've been to.

Perfectoids have been instrumental for yielding quite a few important results in the last few years. Could we have guessed this development would have occurred? I think not. We should be very open to the future of mathematics.

Chapter 11

Interview with Richard Nowakowski

Richard Nowakowski works on graph theory and combinatorial game theory at Dalhousie University, where he is a University Research Professor. His students call him *RJN* for short. He completed his M.Sc. and Ph.D. degrees under the supervision of Richard Guy at the University of Calgary. Richard won the Adrien Pouliot Award in 2007 from the Canadian Mathematical Society for his outstanding contribution to education and outreach.

Richard is retired now, but I have fond memories of him in my office writing the final parts of our joint book *The Game of Cops and Robbers on Graphs*. Loved by his many students and collaborators, Richard has a keen mind, an inextinguishable curiosity, and is serious about play.

This interview was conducted in March 2016, and was the very first one posted on my blog.

∞

AB: How did you first become interested in mathematics?

Figure 1. Richard Nowakowski. (Photo courtesy of Dalhousie University.)

RJN: I am one of those people who was always interested in mathematics. In grade two or three, I was winning gold stars for being able to add up columns of three-digit numbers much faster than anyone else. So I received reinforcement very early on.

Anytime anything mathematical came along I was just interested in doing it. Just like sports where you go out and develop your talent, I just developed a little more mathematical talent by being more aware and interested in doing those little things. One of my aunts was a teacher, and we had this 11-plus exam in England. I was at a small school; there were six out of thirty who might have a chance of passing the exam. She got us all sorts of interesting books to study. I wasn't good on the English side, but the math ones I found interesting.

Ages eleven to fifteen, I was doing eight to ten subjects each year in school. Sometimes teachers were not math teachers, but they still taught math. I had enough confidence with friends to essentially ignore their teaching to take the content and work with that.

When I immigrated to Canada in 1968, I was sixteen, and they put me in grade 11 although I could have gone straight to university with my academic background. I went from an all-boys school to a co-ed school. There were more distractions! A new country, new social scheme, everything. But I enjoyed it.

In first year university, I almost failed calculus because I missed the midterm playing Bridge instead. However, I saw all the material two to three years earlier in England, and I had close to one hundred percent on the final and assignments so passed the course.

There was a new guy, Jedrzej Sniatycki, at the University of Calgary who was given the task of teaching a new course called Sets and Mappings. There were seven of us in that course, who ended up being Honors Mathematics and Honors Computer Science. We spent a lot of time together in the next four years. We took the course and had fun with it. He did things like the different kinds of infinities of sets which was new to me. Sniatycki wasn't bound by doing the straight proofs. He had a lot of fun with the material, and we did as well.

I thought I would study either mathematics or physics, maybe even astronomy. Physics bored me though, and they weren't doing anything new in calculus. The Sets and Mappings course was new math that I hadn't seen before.

AB: Who influenced you to study mathematics as a student?

RJN: I took Sets and Mappings in first year. In second year, the guy teaching abstract algebra did some game theory like Nim. John Conway was visiting Richard Guy at Calgary, who I didn't know at the time. I met Conway, and I still remember the first question he asked me: "What is $1 + 1 + 1$?" I answered "3", and he replied "Oh you are so naive!" Of course, he was talking about combinatorial games.

At that time, I became aware of Richard Guy, mainly because of the noise his machines would make. He was running an electromechanical computing machine, which ran in base ten rather than base two. It ran 24/7, was loud, and had moving parts like ball bearings. About once a month the machine would let out this high pitch whine or scream. I would be walking down the hall past his office, hear that scream, and wondering what he was doing.

I had a number theory class with Guy. He wowed the honors students in that class. He came in one day talking about continued fractions for π. He first wrote out π to thirty decimal places from memory on the board and wrote the lecture with no notes. I was sitting in the back and was wowed.

Guy's assignments had some easy questions, a few interesting ones, and always one final one where I was clueless. My classmates and I would meet the night before over pizza, having completed all but the last question. We had no idea how to do the last question. That was a lot of fun.

AB: Was the final question like a research problem?

RJN: Some of it could be research, and some was extreme problem-solving. At the end of one of his 8 am classes, I told Guy that the last question on the latest assignment was tough. He said it had a solution that was one line, but I wrote two pages in my assignment. I figured I did it wrong. This was Wednesday and the next Monday he came back and said that we were going to write a paper together based on my solution. I misunderstood the question but in an interesting way. It did appear as a paper several months later, and it is still the longest and best math review of any of my papers.

Again, it is this notion of inclusion. There are many mathematical questions that no one has thought about, and even undergraduates can have a good idea about these things. In my fourth year, a new guy came along, Ivan Rival, who was definitely a type-A personality. He was also inclusive. His approach in teaching was that while he knew more than his students, that was due to experience. He would present problems, and we would talk and think about them. Rival would work with anyone. So even though he wasn't my supervisor for either my Masters or Ph.D., up until recently, he was my most prolific co-author.

I wasn't planning on becoming an academic, but then I got an NSERC scholarship. I thought Guy was doing interesting things, and

I wasn't looking for a job, so I worked with him for my Masters, and with Ivan on the side on partially ordered sets and combinatorial number theory.

Figure 2. Richard K. Guy. (Photo licensed under Creative Commons Attribution 2.0 Generic (CC BY 2.0).)

Guy said that if I stay with him for a Ph.D., then he would take me to Cambridge for a year. I took him up on it. In my second year, I went to Cambridge, and again the atmosphere there was centered on inclusion and playfulness. Conway was there, along with others; Guy was visiting Conway, and they worked on their book *Winning Ways*. Someone would say here is a new problem and everyone played and did things.

AB: Would you explain the role of games in your mathematical work?

RJN: I would separate out play and games. Mathematical play was what Guy, Rival, Conway, and his cadre of people did a lot. Inclusive play. Stretching yourself. You don't know where the next idea is going to come from.

I got interested in games by watching what they were doing. I would say that is separate from play. I didn't do much with combinatorial game theory until the late 1990s. Quickly after I got to Dalhousie, I was classified as a graph theorist and taught the subject. I had a couple of interesting questions early on.

George Gabor (a Hungarian statistician working at Dal), asked me the question that became Cops and Robbers. As I thought about it, the technique for characterizing cop-win graphs came to me, and it was identical in form to a problem from partially ordered sets on fixpoints. It was like two different questions, but the same technique. It was clear this question was going to set me off in a different direction. I found that very, very intriguing. I tried to roll the two ideas together over the next few years. As my graduate students will tell you, I did so spectacularly unsuccessfully since all my conjectures were false. But it was good for the graduate students, and they could tell me I was wrong.

AB: What are some of your favorite results?

RJN: I would mention the characterization of cop-win graphs, where one cop can capture the robber.

The one I also enjoyed was the characterization of well-covered graphs of girth five or greater. These are graphs where the smallest cycle is a 5-cycle, and you consider independent sets of vertices, where no two are adjacent. For instance, think of fire stations: we don't want them to be too close. Well-covered graphs are those where every maximal set of these independent vertices are all the same size. This makes it easy to find one, as it is a hard problem to decide which is the biggest one in general. This was joint work with Bert Hartnell and Art Finbow who were on sabbatical at Dalhousie University.

One of the reasons I liked this problem was because of the adrenaline rush. For a few months, we couldn't get any results. My subconscious was trying to tell my conscious mind that there was something there. It doesn't always work that the subconscious is correct. After

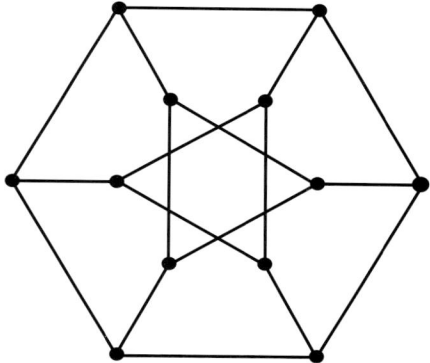

Figure 3. A well-covered graph. Every maximal independent set has size four.

three months, every single idea was incorrect. Finally, we came to the solution. It was an odd time in a way, but it was a lot of fun to finally finish.

I enjoyed most of my research projects, especially working with other people. One of the strangest ones was working with Ivan Rival on a problem of his on partial orders. He thought in terms of formulas, while I am very visual. I said "Do you mean the following?" drawing something rough on the board. He looked at it for thirty seconds and said: "Do you mean the following?" I asked the same again, and we went back-and-forth like this for four hours. In the end, we had a paper! It wasn't a major result, but it was interesting to see how two people could do this back-and-forth. It was like playing tennis.

AB: Do you think of mathematics as science or art or both or neither?

RJN: I think the perception in the discipline has changed over my lifetime. Early on it was something you did in your office by yourself. In that sense, it is more like an art. We have moved on to more collaboration, which requires a slightly different mindset. But now with computers, we can find new things. We can get intuition from computers that we could never get before. So, we are moving from the all-artistic model of mathematics to a mix of the art and the science. I think the art is still important... we are not doing this for the money that's for sure!

So there has to be some internal satisfaction that you get from the work. You have to be doing something you enjoy, which is again on the artistic side. You might have a question that will create lots of papers and attention, but it's totally boring as far as you are concerned.

I think the artistic and creative sides are important. You keep working on it, it keeps your interest, and you find those little niches. You will get a lot of satisfaction.

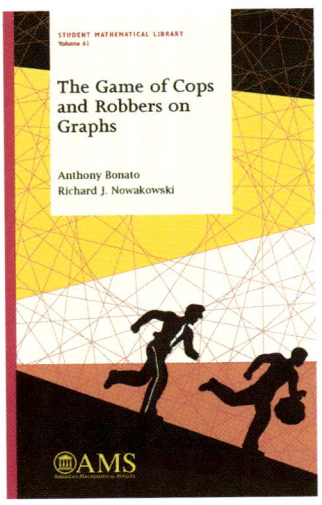

Figure 4. Richard's book on Cops and Robbers, joint with the author, published by the AMS in 2011.

AB: What role do students have in your research?

RJN: I took the model of Richard Guy and Ivan Rival. I find it a lot of fun to work with people. I do my best work with others, and most of my work is with co-authors. Graduate students don't have as much experience or background, but they do have a lot of raw talent. We are never quite sure where the ideas will come from. It is a way to keep me young, I suppose. Seeing them take off is great.

Through my tenure as Chair of the department, it was necessary to have good graduate students to keep me coming out the other end as a sane research mathematician. I think students are very important and the best ones are self-motivated. They have to have confidence, not necessarily a great background. They need the confidence to stand up to me and say I'm wrong. It's one of the reasons I like games; if you play a game by yourself, it gets boring.

AB: Do you have any other interests besides mathematics?

RJN: My big three outside academia are: playing Go; Bridge (which I gave up after that incident with the calculus midterm); and Chess (years ago back in my school team; I don't play now).

Figure 5. The game of Go. (Photo from Shutterstock.com.)

I also like international folk dance and English and country dance. You find a reasonable number of scientists and mathematicians who do that, and I am not sure what it says. My other two hobbies are hiking and gardening.

AB: What do you think are some of the important directions for the future of mathematics?

RJN: That's a tough one. I focused my career on discrete mathematics. What I am seeing there is a move to more dynamic questions like Cops and Robbers and graph cleaning. Also, things with probabilistic approaches are becoming more important. I am not so good at those approaches, but I can see what it means and how it changes the way we look at problems.

I think someone going into discrete mathematics now needs a well-rounded education: for instance, it would be good to have a background in algebra and the probabilistic method. Programming skills are very helpful. Being able to use those is important since they can give the intuition on the problems where pen and paper aren't enough.

Even if you don't have the right background, if you are well-rounded then you can find the right people to ask. Right now, in my work on combinatorial games, we are doing some new stuff on scoring. We have to ask people in category theory and algebra for help. To algebraists, no one has looked at these particular structures since the 1960s. But we are asking questions no one asked before in this area. If one of us didn't have algebra in our background, we would be doing a disservice to mathematics by putting out incomprehensible stuff.

Chapter 12

Interview with Ken Ono

Ken Ono is the Asa Griggs Candler Professor of Mathematics and Computer Science at Emory University. His highly accomplished research program focuses on number theory, algebra, and combinatorics. Ken graduated from UCLA with his doctorate of mathematics in 1993. Before joining Emory in 2010, he has held several academic positions including one at the Institute for Advanced Study at Princeton University.

Figure 1. Ken Ono. (Photo courtesy of Ken Ono.)

Ken received great recognition for his scholarship. His many awards and recognitions include an Alfred P. Sloan Foundation Research Fellowship, a Packard Fellowship, a Guggenheim Foundation Fellowship, the Presidential Early Career Award, National Science Foundation grants, and the National Science Foundation Director's Distinguished Teaching Scholar Award. He is also a Fellow of the American Mathematical Society.

Apart from his mathematical work, Ken has been active in various media productions. Most notably he was an associate producer and consultant for the movie *The Man Who Knew Infinity* based on the life of Ramanujan.

Ken has an engaging life story, which was highly influenced by the legendary mathematical genius Srinivasa Ramanujan. It's not every day you meet a top mathematician who was a high school dropout. There are some wonderful life lessons there for all of us here.

This interview was conducted in September 2016.

$$\infty$$

AB: When did you first realize you had a talent and interest in mathematics?

KO: Those are two different questions. I recognized that I had math talent in first or second grade. My parents like to tell the story that I independently discovered the proof that there are infinitely many primes when I was in second grade. I can't vouch for that.

For as long as I can remember, I was told I was good at mathematics. I didn't love mathematics until my early twenties.

AB: In your recent book *My Search for Ramanujan* you chronicle the struggles that led you to a career in mathematics. For those who haven't read the book, would you describe some of the background leading up to your doctoral work in mathematics?

KO: My parents came to the United States in the 1950s. My father got a job as a member at the Institute for Advanced Study. André

Weil invited him. Try to imagine their circumstances. My parents, both Japanese, were brought up in an environment where they were told that their emperor was a god. They were told that their race was superior to the Americans. It was difficult for all Japanese at the time to face the reality that this is absolutely false. How do people like my parents recover from all that?

My parents were not embraced when they came to the US. In fact, there were terrible things that happened to them because they were Japanese—they were considered the enemy. And it was difficult for them to raise three boys in a country they were taught to hate. My parents thought that the only way for us to succeed was for us to be great mathematicians or scientists or musicians. They prohibited us from doing anything that was normal. They were Über-Tiger parents.

My book begins with how it was difficult to live under these circumstances. I grew to hate math. I knew I was good at it, but by the time I was in high school, I hated math as it represented everything I did not want to be. It represented not having freedom. I couldn't hang out with friends. I wasn't even allowed to have friends.

It was a very long road. A large part of my book is about coming to terms with my parents. I love my parents. I only understand them now after having spent a lot of time imagining what it would have been like to be them, raising three boys in a country they were afraid to live in. They did their best. However, opinions changed over time. We were no longer discriminated against by the time I was in grade school. The atrocities my parents suffered when they first came to the US were so extreme that even to this day they lock themselves in their house. I'm the youngest of three boys. By the time I was in elementary school, I didn't know about any of this. All I knew was my life with my brothers in our small nuclear family. We were on our own, locked in our house in isolationist Japan going to school in America.

Ramanujan helped me at basically every critical point in my life. I am a high school dropout. In the 10th grade, I was planning on dropping out of high school to run away from home and never see my parents again—that's how stifling it was. On April 7, 1984, when I was in tenth grade, a letter came to the house from India. It was from an Indian woman I never heard of called Janaki Ammal (who was Ramanujan's widow). She thanked my dad, who was one of eighty mathematicians who contributed to help pay for the statue of some Indian guy who died in 1920. He broke down after reading this letter. He then told me the story of this legendary mathematician Ramanujan whose ideas came to him in visions from a goddess.

All I could hear was that my dad was looking up to a two-time college dropout as a hero. Maybe if I don't get perfect college admissions test scores, they wouldn't kick me out of the house! Maybe there was a path for reconciliation.

So, I used the letter against my parents saying it was unreasonable for them to demand their kids to be perfect students. I threw Ramanujan back in their faces thinking it would be a source of more arguments. To my surprise, it worked. He said if you need time off (and this was before anyone talked about gap years), then drop out of high school and go find yourself. But he asked to do it safely, and go live with your brother Santa who is a graduate student in Montréal. A few weeks later, I was living with my brother in Montréal enjoying my freedom. I never understood why my dad let me go, until 1997. I went to the University of Chicago. It was easier to get into college then than it is now. Chicago had a program that admitted kids without high school diplomas. I had test scores off the charts. A psychologist at Johns Hopkins had been studying me since I was five years old. His name was Julian Stanley. He was very famous. He wrote a letter of recommendation for me that got me into the University of Chicago. Without that, I don't know what would have become of me.

I was a terrible student at the university, majoring in Greek life. By the time I was a senior, I was struggling with a 2.7 GPA. I wanted to be pre-med—that didn't last long. I was lost. In the summer of 1988, I was flipping through the channels on television and came upon a NOVA special on Ramanujan. I hadn't thought about him in four years. On television, in color, was the story of Ramanujan. It transfixed me. Had it not been for that accidental evening flipping channels, I might have worked for a bank. Instead, I thought about what Ramanujan did for me as a tenth grader, and I found hope again in his story. After much soul-searching, I decided to make the most of my very last year of college.

Beforehand, I didn't go to class, and, if I did go, I sat in the back row and wanted to be anonymous. That was hard to do, as my father was a math professor and had many friends at the University of Chicago. I did the minimal amount of work to pass.

I took one class from Paul Sally, which was a graduate class in analysis. I tried and did very well. By the end of the first quarter of my senior year, he called me into his office and said that he wanted to make a covenant with me as he had figured me out. Paul Sally had an interesting story of his own: he started as a taxi cab driver. He took Ray Kunze at Brandeis University to the Boston Logan airport, and that was how he was discovered. When he was my age he wanted to play basketball. He said I was talented, and that I should go to graduate school. We made a deal, and Sally made some calls. He got me into top graduate programs despite my 2.7 GPA.

In graduate school at UCLA, I had a great advisor, Basil Gordon. We read *The Man Who Knew Infinity* together. I became a mathematician because Basil taught me to see beauty in mathematics and I haven't turned back since.

Ramanujan helped me seek my own life at sixteen, and when I was in college, it woke me up to the fact that I should forge a life for myself, not just continue to be angry with my parents. As a professional mathematician, I have been following some of the paths he has left behind.

Ramanujan was an unexpected guardian angel. Every time he entered my life, something amazing happened. And most recently there was the film *The Man Who Knew Infinity*. If Matt Brown hadn't made this film, my life would have been different.

AB: You were a consultant on the excellent Ramanujan biopic *The Man Who Knew Infinity*. How did you get involved in the film? What was your experience working on the movie?

KO: It was June 2014 and I got an e-mail from Matt Brown about his film project *The Man Who Knew Infinity*. They were in Pinewood

Studios getting ready for filming. The next day I spoke with Liz Colbert who was the art director for the film, and she wanted to talk about the props. What was supposed to be a fifteen-minute introduction turned into a three-hour chat. Matt realized I knew the story so well that I could be useful on set.

Figure 2. Ken Ono at the world premier of *The Man Who Knew Infinity* at the 2015 Toronto Film Festival. (Photo licensed under the Creative Commons Attribution-ShareAlike 4.0 International (CC BY-SA 4.0).)

A few days later I flew to London not only to help with the art but the final tweaking of the script. I also helped with the rehearsals with Dev Patel and Jeremy Irons. I ended up being promoted to

Associate Producer. Now I'm in the film business, and I'm having the time of my life.

AB: There is a great photo of you explaining mathematical concepts to Dev Patel on the set of *The Man Who Knew Infinity*. What was that experience like? Do you think he grasped the mathematics involved?

KO: We were rehearsing scenes on Hardy's work with Ramanujan on the partition function, where they were trying to obtain an asymptotic formula. If you asked him today, Dev could probably recall the partition numbers, and that they grow incredibly fast. He could also tell you that Ramanujan and Hardy figured out a formula that nearly gives the correct answer. Anything beyond that he didn't understand, but time was limited. Our goal was to get the spirit of the work right. And I think he did a great job.

AB: There is a burgeoning interest among the public to watch movies and read biographies about theoreticians like Ramanujan, Hardy, Turing, Hawking, and so on. Why do you think the people are interested in mathematicians and their stories?

KO: There are two projects in the works involving Nicolai Tesla and Emmy Noether. A good screenwriter can make a movie out of any interesting person. With the success of *A Beautiful Mind* and *Good Will Hunting*, the idea of making movies about mathematicians was in the air.

You have to have producers and financiers who are willing to underwrite these projects. That was the main obstacle for decades. Great actors are also trying to put capstones in their careers. That's how we got Jeremy Irons. He's played Alfred the butler, and the priest in *The Mission*, but he wanted to play the role of a Cambridge don. Hardy was perfect for the role.

AB: Your research focuses on a number of topics such as number theory, combinatorics, and algebra. In broad terms, can you describe the goals of your research program?

KO: Early in my career I focused on number theory. I focused mainly on classical number theory, like the theory of partitions that Ramanujan studied. I was also interested in elliptic curves and the Birch and Swinnerton-Dyer conjecture. I haven't thought about those topics in a long time, although many of my students still work in those areas. More recently I've been thinking about the objects I study in number theory, and their role in representation theory, string theory, and mathematical physics. We recently proved the Umbral Moonshine Conjecture, which is at the interface of number theory, physics, and representation theory. Most of my work is now related to these questions.

Earlier this year with Kannan Soundararajan at Stanford and Seokho Jin and Wenjun Ma, we proved the Riemann Hypothesis for modular form periods, and we are proud of that result. Most of my time now is focused on moonshine and representation theory.

AB: How did you come to solve the Umbral Moonshine Conjecture?

KO: Here is Ramanujan again. It's unbelievable.

John Conway and Simon Norton noticed decades ago that the coefficients of the modular j function seemed to resemble the degrees of the irreducible representations of the monster group (the largest of the finite simple sporadic groups). This came at the time before the completion of the classification of the finite simple groups.

Conway and Norton assembled a precise theory which showed that the representation theory of the monster group is completely encoded by a special class of modular functions. Later, Richard Borcherds proved this conjecture by constructing an infinite-dimensional monster Lie algebra that makes all of this precise. He went on to win the Fields Medal.

Ten years later the mathematical physicists, in working with the K3 elliptic genus, had a set of coefficients to study just like Conway and Norton had for the j function. They discovered that their coefficients were in sync with the irreducible representations of the Mathieu group M24, which is another one of the larger sporadic simple finite groups. The functions they studied weren't modular but were mock theta functions. No one knew what a mock theta function was until 12 years ago. Ramanujan coined the term in the last letter he sent to Hardy from his death-bed (three months before he died).

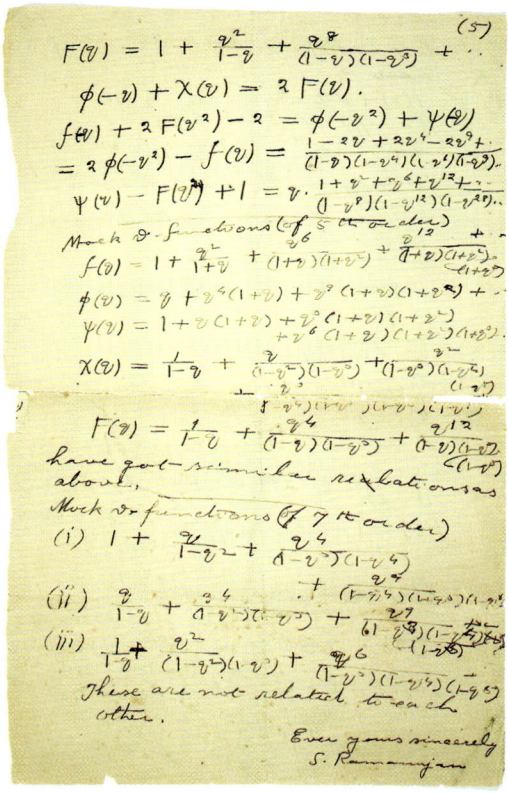

Figure 3. A page from Ramanujan's last letter to Hardy, where he describes mock theta functions. (Photo courtesy of Ken Ono.)

I had been studying the mock theta functions for six or seven years when the Japanese physicists made this observation, so I started to pay attention. Three years ago, at the Simons Center in Stony-brook, there was a whole semester devoted to this area of string theory, and I gave a lecture on the opening workshop, as did the researchers Miranda Cheng, Jeffrey Harvey, and John Duncan. What they did was to build on the observations of the physicists and posed the Umbral Moonshine Conjecture. The conjecture has to do with the representation theory of the automorphism groups of certain lattices which included M24. The conjecture was that the representation theory of all of these groups comes from the mock theta functions that I had been studying. I teamed up with John and one of my former students Michael Griffin, who is now a post-doc at Princeton, and we proved it. That marked my entry into the subject.

I have a paper coming out in a few weeks where we generalized the theory of the monster denominators formula of Borcherds' to every modular curve. Thanks to Ramanujan, the formulas were recognized, and we developed some general theory. We were lucky to be working on these topics just at the right time.

AB: You're a very active supervisor, with twenty-five graduated doctoral students (and counting). What advice do you give your students about working in mathematics? More generally, what advice do you give young people (in high school or even younger)?

KO: I don't talk so much about working in mathematics, but instead I talk about finding your passion. I go out of my way to de-emphasize the role of test scores and grades and memorisation. We live in an era where there is this temptation to brand everything; for example, if you went to Harvard or MIT, you must be smarter than those who attend other schools. Whatever happened to evaluating people for the quality of their achievements and character?

I want my students to call me "Ken". I'd rather be evaluated for the advice I offer rather than a title.

I love talking to young people. It helps keep me young. I am 48, but almost every day I get to go to work and be around these great people with their whole lives in front of them. It's a pleasure to be part of people discovering themselves. It's so great to have jobs like the ones we do. We're lucky.

AB: Would you tell us about the talent search the Spirit of Ramanujan Math Talent Initiative?

KO: In Ramanujan, we have someone who was a role model. He was my role model. I believe fully in his message, which is that talent can be found in the most unpromising of circumstances. Ramanujan was an Indian who grew up at a time when you didn't want to be Indian, and his ideas were so important that we are still talking about them today.

The idea of our global talent search is our common concern of finding a present-day Ramanujan. He was a two-time college dropout who we would ignore today because of poor test scores or grades. In every discipline, there are hundreds, even thousands, of people who are slowly contributing to a field, but it's only once in a while someone like Ramanujan comes along and propels subjects forward. Finding people like Ramanujan is vital.

With help from the Templeton World Charity Foundation and Expii.com (a company founded by my friend Po-Shen Loh who is the International Math Olympiad Team coach for the US, and also trains many Canadian students), we are searching the world on-line through this site for undiscovered math talent. We hope to award a scholarship to twenty or thirty promising young mathematicians. A dozen or so we will link with a mentor. Maybe we will find a present-day Hardy for those who want that relationship.

We already have a few winners. We found a boy in Qatar named Ishwar Karthik who is twelve. When he was ten, he discovered his method for computing digits of π. He was living and working in isolation in a desert. He's now connected with a professor at Texas A&M at their Qatar campus. They meet once a week, and they are having the time of their lives. We found a boy in my hometown of Atlanta named Dean Cureton, and that kid is amazing too. He aced the AP calculus test (an achievement test that pre-college students take) at the age of eleven, after having taught himself calculus from

Figure 4. Srinivasa Ramanujan (1887–1920). (Photo in the public domain.)

the internet. We meet with him every Saturday. He is super smart, and he's attended some of my college classes at Emory. Maybe you will see him on TV someday, as there is interest in having him on the Steven Colbert show and Jimmy Fallon show. He is so enthusiastic, and he loves math.

AB: Maybe you will play the role of Hardy for him?

KO: I can't ever really be Hardy, but it's a treat to find these kids and help them on their way. It's not just the young kids, but it's also about students in college and my summer REU program. It's all about the process.

I benefited from having great mentors like Paul Sally, my advisor Basil Gordon, and later Andrew Granville. I cannot thank them enough for what they have done for me. Andrew is still alive unlike the other two, and he's one of the most influential living analytic number theorists. The way I could honor the memory of Paul Sally and Basil Gordon is to do for others what they did for me. Without them, I don't know where I would be right now. And there is never a day that I don't reflect on that.

Figure 5. Paul Sally (1933–2013). (Wikimedia Commons, public domain.)

It doesn't matter how talented you are; everyone needs help. It could come in the form of a parent, a friend, or others. If you are in the position to offer it, then it's one of the best things you can do to make the world a better place.

AB: I'd like to finish by looking towards the future. What do you think are the major directions in mathematics?

KO: I'm very worried about the future of our field. Not because we are going to run out of questions. I'm very worried about the actual inner workings of our profession. It's not clear what professional publishing will look like in the future. We've been talking about the open access mandates for years, and now there has been some progress that worries me. I'm worried about the continuation of federal funding. I'm absolutely frightened about what the presidential election will bring in a few weeks. I have very high hopes for one candidate and am utterly petrified if the other candidate wins.

There is no denying the world is facing many challenges: political, social, and environmental. Having a scientifically literate citizenship is important, and we are in danger of losing that. The professional societies (for example, the American and Canadian Mathematical Societies) must be very proactive with their public policy committees. I think they have their work cut out for them in the future.

Index